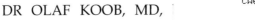

 DR OLAF KOOB, MD, Medicine and worked as a burg and Wanne-Eickel. He has been a drugs counsellor, a general practitioner in Weimar and Berlin, a school doctor at the Berlin Therapy Centre for children with special needs, and spent many years collaborating on a research project into drug-related diseases and social factors. He currently lectures and leads seminars in Germany and around the world and is the author of many popular books on medicine and healing.

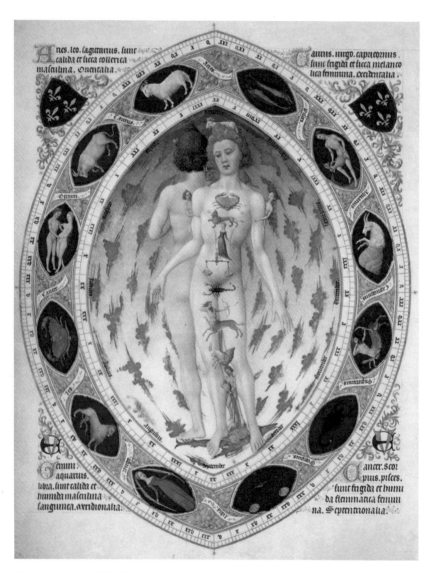

'The Anatomical Zodiac Man' from The Very Rich Hours of the Duke of Berry

IF THE ORGANS COULD SPEAK

The Foundations of Physical and Mental Health

Understanding the Character of our Inner Anatomy

Olaf Koob

TEMPLE LODGE

Temple Lodge Publishing Ltd.
Hillside House, The Square
Forest Row, RH18 5ES

www.templelodge.com

Published in English by Temple Lodge 2018

Originally published in German under the title *Wenn die Organe sprechen könnten, Grundlagen der leiblich-seelischen Gesundheit* by Info3 Verlag, Frankfurt am Main, 2005

© Info3 Verlag 2005
This translation © Temple Lodge Publishing 2018

Original translation by Pauline Wehrle reworked by Temple Lodge Publishing

A CIP catalogue record for this book is available from the British Library

ISBN 978 1 912230 15 0

Cover by Morgan Creative featuring detail of a self-portrait by Albrecht Dürer
Typeset by DP Photosetting, Neath, West Glamorgan
Printed and bound by 4Edge Ltd., Essex

Contents

We know just about nothing about our bodies. We have been brought up the wrong way. I can actually telephone to San Francisco from my own room, whereas I have no idea at this moment about what is going on in my liver or in my gall bladder. It will be the duty of modern education to make us conscious of the inner organs. I could and should get to know about the functions of my stomach, glands and bladder just as well as I know of the movements of my hands, eyes and mouth. We human beings are, as regards ourselves, still not strong and brave enough. Human beings already dare to look the stars in the eye, but where their spleen or their intestines are concerned they still lack the courage to do so. We should be acquiring a deeper self-awareness, some sort of x-ray-like connection with ourselves. But this path is more difficult and more eerie than a trip to the moon.[1]

Sandor Marai

Preface

Our western world, that appears to be so enlightened, behaves with regard to many practical questions that are vital to life in an extraordinarily paradoxical way. Two examples may make this clear: firstly, the ignoring of old age and its fruits acquired over a lifetime, along with the illusion of eternal youth. The ratio of young people to old people in our society is, though, shifting dramatically in favour of age. Whilst celebrating the ever-rising life expectancy underlying this, the problem of caring for the old, now almost insoluble in our society, is overlooked. In short, people want to grow very old, but on no account *be* old!

Medicine can serve as the second example. Since the middle of the nineteenth century rational scientific medicine has been making every effort imaginable to get rid of illness. We supposedly understand the causes of illness and are developing the means to overcome it, primarily by means of more and more effective and above all more expensive remedies. This 'health market', with its services to society, is now a very important pillar of public expenditure. It has the greatest economic potential and fastest rate of growth, even at times of recession. This market would completely collapse if its remedies really brought about a cure and drove illness away. For then we would not need them any more!

A third paradox is of particular concern in this book. The human physical body is valued today more than it ever was before. Body-care cosmetics, ways to improve the body with supplementary remedies, dietary foods or medicaments, changes of body-shape by means of cosmetic surgery, and many other such ways in which the body is altered and affected, supply a wide market that, just like the health market, is a significant part of a prospering economy. Yet although so much value is placed on making the body look good and function well, so that, ideally, we are scarcely even aware of it any longer, we understand virtually nothing about this body today, its functions, or the developmental and formative principles underlying our organism. Even apparently cultured people or those who are highly esteemed in society will soon admit that they find it difficult to say where the adrenal glands are, what we need a thyroid gland for, how a liver is organically structured and a lot more. Whereas human beings increasingly want to determine matters for themselves in their democratic social systems, and whereas we will, it seems, fight tooth and nail for our personal freedom, there is an almost incomprehensible lack of understanding of how our own body and its individual organs and organ systems such as kidneys, bones and muscles or blood are built up and function, how their co-operation is organized, how they are more or less capable of regenerating themselves, what helps them and what harms them. Where questions such as these are concerned, an otherwise enlightened person has returned to the Stone Age.

Olaf Koob wants to change this. With the vision and experience of a holistic physician, he describes the world of the organs in their harmony as the body, soul and spirit of an individual person. He trained originally as an orthodox doctor, learning to think analytically and studying the whole in its parts. His method of presentation is, however, one of synthesis, describing the parts as the expression of a whole. And the whole is more than the sum of its parts! Much in the organism only becomes clear for instance through the organs' collaborative or also contrary actions, through the action of very diverse soul forces, and through increasing individualization during the course of life. The author offers a survey of the most diverse medical systems, and describes their common essence. Orthodox medicine, naturopathy, homeopathy, Ayurvedic medicine, Chinese medicine and anthroposophic medicine are so familiar to him that he works out of what they have in common and what is special to each. So the organs really begin to tell their story, and it is up to the reader first to learn to hear it and then to learn to understand it.

This is not a formulaic or programmatic book, nor a reference book, but one to pick up to read again and again. At a time in which medicine is passing through great change, because an understanding of illnesses (pathogenesis) is slowly being replaced by the understanding of health (salutogenesis), and individuals are increasingly called upon to be responsible for nurturing their own health, it is indispensable to gain ever greater insight into one's own body as a living and feeling—and in other words ensouled—organism, and to learn what specifically and individually helps it or, on the other hand, harms it. Explanatory presentations by a physician are essential here. This book can help us trace the secrets of our own body, to see it as a wonder of creation, to marvel at it time and again with reverence and gratitude, because it is seeks to serve us, selflessly, for a whole lifetime.

Prof. Dr Volker Fintelmann June 2005

Foreword

To assuage the imaginings of a sick person and reassure him, at least, that
he does not have to suffer as before more from his thoughts about his
illness than from the illness itself – I think that really is something! And it
is not a mere trifle.[2]

Friedrich Nietzsche

In a book intriguingly entitled *The Mysteries Within: A Surgeon*
Explores Myth, Medicine, and the Human Body,[3] a well-known
American surgeon and medical historian describes his experience over decades in dealing with certain organs whose organic
'intelligence' and relationship to soul life are still a great mystery
to us, although, in purely technical terms, we can control their
functions and even replace them. Why, for instance, does a liver
that has had part of it removed, grow again to a considerable
extent, and why does this not happen with lungs or the heart, let
alone with the brain? Why, despite all technical progress, do we
still know so little about the task of the spleen, an organ that one
can even entirely remove leaving the body relatively undam-
aged? It was not for nothing that the great Roman doctor, Galen,
called the spleen the *'organum plenum mysterii'*, the 'mysterious
organ'. And what makes our heart into such a special organ that,
for thousands of years, humanity has not tired of writing paeans
and poems of praise to it, although it is still regarded more or less
as a 'pump' and is even transplanted?

In the course of his observations of organs, with simultaneous
references to the past history of medicine, the above-mentioned
author, Sherwin Nuland, describes a meeting with a Chinese
colleague during a medical conference. Naturally this conversation was primarily about the different views of western and
eastern medicine. Eastern traditional medicine is based on a
worldview of wholeness and quality, whilst the western-
oriented view of science is dominated by a quantitative, analytic

view. At present they are irreconcilably opposed to each another, and therefore need to be reconciled in a way that takes the physically perceptible side of life just as seriously as the invisible and immeasurable nature of the soul-spiritual realm.

In the course of their professional discussion, the Chinese doctor who was also knowledgeable about western medicine said of traditional Chinese healers: 'When they talk about the liver or the thyroid, for instance, they don't mean the organ itself. What they are referring to is the *idea* of the organ. That's a very different thing. For that, anatomy is not needed.'[4]

If therefore we do not regard specific physical phenomena and activities within the organs as the fundamental expression of a living and ensouled whole that finds its visible expression in bodily symptomatology, and if this whole is not rediscovered in every single organic structure—if you like, 'heaven' on and within the 'earth'—then it is scarcely possible to build a bridge between them. That, incidentally, was also Goethe's wish for medicine when he jokingly but aptly said: 'A doctor must be fit for anything. We began with the stars and ended with chicken's eyes.'

We think that a reconciliation of these two opposing attitudes can only succeed when separate things, which for scientific reasons have to be isolated, are again connected up with the superordinate whole, and on the other hand the comprehensive idea, as 'spiritual bond', also makes the distinct and separate parts explicable. In other words we must be able to trace the 'idea of the human being' right into organ functions of the liver, gall bladder, heart and kidney, otherwise the liver is just a liver, the kidney just a kidney, apparently the same as the organs of the same names in cows or pigs.

A great many details of the anatomy and physiology of individual organs are taught in medical studies, but has anyone ever understood from their training anything about the *essence* of the liver—which the Chinese doctor calls 'the general', the Russians speak of as '*pechen*', from '*pech*' meaning an oven, or which ancient Greek and medieval physicians connected with the planet Jupiter? But once we gain a holistic, overall idea, this cannot become or remain vaguely mystical, but we must endeavour to make the various details of the organs qualitatively comprehensible: the position, colour, form, embryonic development of each. Then the living biography and physiognomy of an organ or even of an illness can be understood in far greater detail. Only then will we really have a holistic medicine worthy of the name!

So we must at least adopt the view of 'The Little Prince' of St. Exupéry and not always see only the outer side of something, see 'the hat' and theorize into it everything we possibly can, but also see everything that is really hidden in the hat-shaped snake — the 'elephant' that is to begin with invisible to the physical eye, the parable-like nature behind all that is transitory.

It seems to us that the bridge we seek can be built by the method that we call the phenomenological imagination. This tries to understand the physical appearance of something as the perceived 'exterior' of a superordinate spiritual dynamic. For without underlying ideas, nothing whatever can arise in the visible world. We need only think here of technology. Why should it be different in the realm of life? In this way organs can become individual 'destiny portraits' with their own past, present and future, as is true of individual people and of humanity itself as it changes over the ages. The exclusively brain-oriented attitude of people is the result of recent development. For instance in earlier times the heart was seen as an organ of destiny and of memory. This can be clearly seen in the English or the French language, when being able to remember something is brought into connection with the heart: 'to learn by heart' or *'apprendre par cœur'*. In the Chinese language the sign for 'thinking' is the picture of the heart.

This book was written for the kind of people who, as laymen interested in medicine, want to extend their knowledge, and do not always want to be told what to think by 'experts'. At a time of increasing uncertainty in healthcare policy, driven also by a pharmaceutical industry founded on perpetuating illness,[5] it is more necessary than ever for people to exercise their healthy human reason. Thus the more people know about their organism, the better they can preserve their health and cope with crises, suffering, fear and pain.

So we will return here, once again, to the ancient ideas of Greek dietetics and hygiene, and the Arabian and medieval rules for health (*regimen sanitatis*); to the intentions of Hufelands as formulated in his 'macrobiotics' — or the 'art of prolonging human life'; to Baron von Feuchtersleben's inspired efforts in his work, *On the Dietetics of the Soul*. Chinese medicine, too, will be accorded its place; as will, not least, the aims of anthroposophically extended spiritual science. While 'salutogenesis' — the theory of health — is increasingly acknowledged today, we should be conscious of the fact that this outlook is by no means a modern invention but much more a rediscovery of ancient knowledge.

How topical an ancient holistic and cosmic knowledge of the human being really is in today's health debate, is apparent from a statement in the *Yellow Emperor*, a Chinese textbook on medicine written around 3,000 years BC. This tells us, in particular, how we can learn what governs the universe and nature, so as to keep our lives healthy and to develop individually. Somewhat resignedly it also informs us that while the 'Dao', the universal cosmic law, was always obeyed by the wise, and even subscribed to by the ignorant, the latter usually did not follow it in practice. Everything opposed to the harmony of nature, it says, must lead to havoc of body and soul.

> This was why the wise did not treat those who were already ill, but restricted themselves to instructing those who were still in good health. They wanted in fact only to teach those who did not offer any resistance to nature [...]. To administer medicaments for illnesses that have already occurred and to suppress symptoms that have already broken out can be compared to the behaviour of people who only dig a hole for a well after they have become thirsty; such behaviour is comparable with not producing weapons until after the battle has already broken out.[6]

The author feels connected with and indebted to this attitude, one held for more than five thousand years, and hopes to find open-minded readers and perspicacious laymen for what is presented in the following pages.

It is not possible in this context to characterize the anatomical and physiological details and functions of the organs in greater detail. There is literature in the Appendix that leads both further and deeper into these matters. The concern here in the first place is to give a 'picture' of what is going on invisibly and visibly in our bodily nature, and of which, although we inhabit our bodies for many decades, we often have less of an idea about than we have about our car or washing machine.

In this respect we feel closely connected with a concern of the French novelist, Guy de Maupassant who, in the periodical *Gaulois* (1882), formulated it as follows:

> What I want to find in a book is people of flesh and blood: they must be my neighbours, must be people I relate to, must feel the same joys and suffering I experience, and must have a little of myself about them; while I read of them, I want to be continually comparing them with myself.[7]

Foreword to the fourth edition

Heiner Geissler, previously a minister for Youth, Family and Health, recently published a book bearing the Latin title, *Sapere aude!* (Dare to know!) — *Why we Need a New Enlightenment*, which is chiefly concerned with factors underlying economic problems. He urges us to acquire knowledge about life and the contemporary situation, to think about things for ourselves and not accept everything people suggest or assert ... an important theme raised as far back as the eighteenth century in the philosopher Kant's article, 'What is Enlightenment?' His view was that individuals should endeavour, through reflection — that is, by activating their own reasoning powers — to overcome 'their self-imposed immaturity'. The same applies equally in medicine. We might say: Endeavour individually, i.e. autonomously, to shape your life, your health and illness, even your death. In order to do this it is of course necessary to possess knowledge of one's own organism and frame of mind, so that, as the ancient doctrine of dietetics proposed, we lead our life consciously, and, while still healthy, find the information we need to remain so. We find these same intentions again in modern 'salutogenesis', which is concerned with maintaining health and cultivating an individual balance of forces. In this respect the author feels himself connected with Baron von Feuchtersleben's work *On the Dietetics of the Soul* and Hufeland's *Macrobiotics – or the Art of prolonging Life*, two significant works from the nineteenth century that have perpetuated both the ancient Greek as well as the traditional Chinese tradition of active prophylaxis.

The author has, in the course of the years, received much positive response for offering a 'missing link' between textbooks for doctors and popular literature on maintaining health. 'Know thyself', the motto of the temple at Delphi, is something that can still be our ideal today, at every level of existence.

Berlin, January 2014

Foreword to the sixth edition

Following lectures or seminars people often ask me where this book's unusual title came from. Originally I planned to use an expression by Nietzsche for the title – 'Reading the Book of the Body', but my publisher at the time thought this sounded too academic. After pondering long and hard, I woke up with the words, 'If the organs could speak' in my ear, and immediately had the sense that this phrase could not have been invented, that it had been sent to me! It seemed to me that my task, in all humility, was to act as a kind of translator of the secret workings of our organs.

I would here like to thank pioneering anthroposophic colleagues whose words and writings inspired me and enlarged my knowledge. I also want to thank my audiences and patients for their positive response and for sharing their own personal experiences, and for thus eliciting some unusual thoughts in me. My encounter with Asian physicians repeatedly confirmed for me that it is possible to have constructive dialogue between different worldviews and pictures of the human being if one is willing to let go of fixed schemas or dogmas.

The current edition has two additional chapters on the pancreas and on the thyroid and hormone glands. I have also expanded my descriptions of the spleen, heart and liver.

Ascona, May 2016

The Human Enigma

Nothing grows in a place where there is no life capable of feeling, growth, or thought. Feathers grow on birds and change every year; hair grows on an animal and changes every year [...]; grass grows on the meadows and the leaves on the trees, and every year they are largely renewed. So we could say that the soul of the earth is its capacity to grow: its flesh is the earth kingdom, its bones are the successive connections of its rocks, out of which the mountain ranges are composed, its cartilage is the tufa, its blood the water veins; the lake of blood around the heart is the world's sea, whose tides are the waxing and waning of the blood in the arteries, which, for the earth, is the ebb and flow of the sea; and the warmth in the soul of the world is fire that has been poured into the earth, while the soul, the ability to grow, lives in the fires that erupt from the earth as baths and sulphur mines and volcanoes, as at Etna in Sicily, and in many other places.[8]

Leonardo da Vinci

Since early antiquity, and through into the nineteenth century, people have regarded the human being as a combination of everything that surrounds us as nature and the universe. In the smallest space, in our body and soul, are not all the forces bundled up that are visibly spread out in the expanses of nature, in the macrocosm, and have come to expression in the various minerals, plants and animals, in the elements and the stars? In the age of the microchip, with its immense functions in the smallest space, this is not such a strange idea. How many big and small products has the development of technology had to discard in the last hundred years, even reject, to arrive at the present form of electronic 'intelligence' on a minute scale? Likewise how many attempts, sketches and 'drafts' has Creation had to produce in order to arrive, for its part, at its highest goal, the human being? In the Finnish language, the human being is called '*ihminen*', which literally means 'little miracle'.

There is actually great scope in the unusual idea that out there, in nature and the cosmos, I see what has been 'shed' from my own human development, and now see in, say, the liverwort, spleenwort, etc., opal, malachite, in ants and bees, even if these express one-sided connections with my organs and soul capacities, but through such connections, can, among other things, be made into remedies in human medicine.

Although I have a strong suspicion that I am right in thinking these thoughts, I still have not even opened the key in the first door into the secrets of that world which can make the connections explicable to my reasoning powers. So, consequently, I must first of all look into the world that I do have access to in order to fully understand my own inner being, my organs. Not until I find this connection shall I have real access to my inner being and the necessary feeling of coherence as a dweller in that world.

Not very long ago a North American journalist did this on a political level when, living in the USA he decided that, in order to understand anything about the USA one couldn't stay in one's own country, because one only gets illusory caricatures. One has to travel abroad, to find out how present-day America and the American people are reflected in the souls of other nationalities. So he set about his quest for self-knowledge by asking other people about his country and its politics. He went home with an abundance of knowledge, some of it quite shattering.[9]

Don't we also get caricatures when we only consider ourselves, our inner life and our organs? Don't we perhaps find much more insight into our inner being when we learn to study detailed aspects of outer nature? It may of course appear a little absurd at first, to have to look *outwards* into the so-called 'objective realm' to experience something objective about my subjective personal self. To be consistent, shouldn't I do the opposite, and look into my subjective inner being first, to find out about the so-called objective world outside?

A caricature of this is the idea people used to have of taking certain drugs to embark on a journey through their inner universe in order to discover something about the outer cosmos. In this connection, however, we can also quote Schiller, from whom there comes this pertinent saying:

> If you seek what is loftiest, grandest, the plant can teach it to you: what in the plant is without will, undertake with your will — that's it!

A study of a growing plant with its lawful development, its leaf metamorphosis and its connection with warmth, light and soil consistency, and its polarity between root and blossom, can indeed tell us something significant about our own soul and bodily development. If on the other hand I could really see into the world of my inner organs I would, as a microcosmic being, acquire knowledge of the greater world and its hidden blueprint. Just as it was customary in ancient Egypt, I could then connect the individual organs with spiritual beings, that is, gods, and know what the liver and the lungs have to do with the myth of Prometheus, the heart with the sun, the intestines with snakes and toads; or how the ant heap, if I truly understand it, offers me a true picture of my immune system. The ant heap, after all, reacts when a foreign body enters it, with its various guards, messengers, killers and clearers, and its other manifold functions, behaving just as 'intelligently' as the immune system in the blood does. This is why we talk about 'immune memory'. By developing such insights we start to explore how and why remedies from the kingdoms of nature can work. There must be a differentiated reciprocal relationship between the various members of the natural world and human beings that can be recognized and made into medicines capable of remedying organic deficits in an illness and bringing about healing — as in naturopathy, homeopathy, the Bach flower remedies, Chinese and anthroposophic medicine.

Many traditional fairytales describe in a humorous and yet profound way how the higher spiritual development of human beings led them to becoming organically 'defective'. There is a Bulgarian fairytale in which the creation of man is described as follows: God lovingly formed many structures out of clay, but none of them corresponded to the original divine intention, so that he had to reject them time and again. As the day drew to a close, he at last succeeded, and after many attempts to make a 'medium race' that satisfied him, he placed his creation in the sun to dry. But he was so absorbed in his work, however, that he did not notice the devil, who came by secretly, and gazed at the work with envy and jealousy. He wanted to spoil God's work, and secretly bored lots of holes in the human body with his stick, which now, to the great surprise of its creator, was disfigured. To save at least the outer form of his creation, God began laboriously to stuff grasses and herbs into all the holes that had been made, smoothed them down with clay, and made his human model intact again. 'But these very grasses, with which God once filled up and mended holes in the body, can be used to cure many an illness. Since that time these medicinal herbs and grasses have existed.'[10]

Thus it was that the devil, the 'devil incarnate'—to whom, according to the creation story of Moses, we owe the knowledge and therewith the freedom of the divine world of the creator—also incorporated into us our capacity for illness.

In Greek mythology we find a similar motif: Prometheus brings fire from heaven to people who, until then, had been dully dreaming and dependent on other beings: an image of enlightenment and the possibility of transforming earthly matter and thus becoming independent of divine leadership. But the revenge of the immortals is inevitable. They send Pandora, the 'richly gifted one', into the world, with a box full of evil and illnesses, that is opened by Epimetheus, the 'retrospective' thinker, as opposed to his brother Prometheus, the 'fore-thinker'. Spiritual possibilities of development have an inner connection with organic weaknesses—the one is not conceivable without the other, for the spirit is the flame that devours growth. Without the possibility of being ill there is no freedom, and conversely freedom brings the possibility to make mistakes and to create illnesses, as we can experience today on a grand scale. Civilization and tooth decay are reciprocal factors, but surplus vitality would also cause problems for us, since it would hinder the acquisition of consciousness. We often find in this respect the remarkable phenomenon that an organically ill person is far healthier in his soul than someone who is always full of vitality—or the other way round.

What kind of relationship is there between the soul and body if, as the saying goes, I am 'liverish', or someone is a 'person of my kidney', or my heart 'breaks'? What do we mean by a 'mental illness'? Can the spirit fall ill at all, or is it just that the soul no longer sounds harmonious—like a piano with a broken hammer? We have to pursue these questions as far as possible in this context because they have many further ramifications for us.

So we shall try to explain the specific nature of various organs and identify their place within the whole organism of soul and body. On the purely anatomical level they lie unrelated as pebbles beside one another. But on the functional level they are connected in ways that Chinese medicine describes as 'organ families', in which particular 'siblings' have a closer connection to each another than to others, and in which there must of course be 'governance' by 'father and mother'—in this case brain and heart. Western medicine sometimes even speaks of a 'concert' of hormone-secreting glands, or of 'closed loop systems', or of the stomach's 'community overlap'.

As far as the relationship of the soul to specific organs is concerned, neither this, nor equally the influence of our organs on soul life, is easy to fathom. But we have been asking these questions for more than 2,600 years. About 600 BC, a pupil of Socrates was suffering from a headache and asked for a remedy to take away the pain and cure him forever, even though he himself was not prepared to change anything in his life—for instance to combat his stubbornness and rashness. But a Greek philosopher is not as easy-going in this respect as modern medical practice! The pupil had to learn from his master that there should not be specialists either for the body or the soul, because they both belong inseparably together; that for instance one could neither understand nor treat a single organ like an eye if one didn't understand anything about the whole, for the eye is part of the head and the head again belongs to the rest of the body, and this, as long as one lives, is gifted with a soul. If the head already produces such problems we can ask if it is perhaps only mirroring something organically and psychologically out of order, rather than itself being the cause.

In fact Leonardo da Vinci also says:

> Anyone who wants to see how the soul lives in the body should also see how this body uses its everyday dwelling; that is, if disorder and confusion reign there, then the body is being looked after by its soul in a disorderly and confusing manner.[11]

But what then are suitable remedies, and if these don't have a good effect, what other, more effective methods could there be?

In an era when we idolize youth, beauty, extending life, and freedom from pain, it is absolutely essential to think about the profounder significance, the meaning, as it were, of suffering, pain, illness and dying. Suffering and pain can, namely, deepen and extend our knowledge of the world, because they awaken a heightened consciousness and by doing so become real 'helpers in our development'. According to Nietzsche,

> the human being, the bravest and longest-suffering animal, doesn't reject suffering as such, but actually *wants* it, goes in search of it, provided one gives him the *reason* for it, tells him *what it is for*. The senselessness of pain, not pain itself, was the curse that has hitherto plagued humanity.[12]

Therefore let us bravely endeavour to find the first of the many mysterious keys to understanding human nature better, and in doing so let us not forget the bright light of reason and our hopefully still sound human understanding. Let us turn away from the smoking lamp of largely habitual prejudices, and continually analyse our own experiences. Let us not unquestioningly believe the self-appointed 'popes' and 'cardinals' of modern science, for 'whoever cites an authority in a dispute is not using his mind but more likely his memory'.[13] Let us avoid that at all costs.

The Cosmic Language of the Human Form

*Whoever engrosses himself in a real way in what we call, in a
Pythagorean sense, studying numbers, learns from this symbolism of
numbers to understand both life and the world.*[14]

Rudolf Steiner

If we want to forge for ourselves the key to the door we must
open in order to recognize that the human being is the result of
cosmic activity, we must try on the one hand to integrate the
human being again into the world as a whole, and on the other
hand to rid ourselves of abstract intellectualism and theoretical
materialism. Thus we can arrive at the kind of sound view that
Leonardo da Vinci stipulated for any scientist and artist: 'All our
knowledge begins with the senses.'

But in order to relate sense reality to its corresponding context,
we require the kind of thinking that accords with our subject, a
thinking that enables us to decipher our many diverse perceptions.

With the 'architectural blueprint' of our body in mind, let us
begin with the apparently most abstract of things, the numbers
and their relationship to one another as the 'end of God's ways'.
As for a house or a temple, an overall plan must exist that enables
the human individuality to be embodied in its earthly 'dwelling'.
It is therefore understandable that in olden times the body was
called the 'temple of the gods'.

If we want to experience something of the beginning of our
existence, we come upon an originally uniform world in which
heaven and earth were not yet divided from one another. We
literally still lay in the lap of the godhead. Accordingly the
number one always represents divine Creation at rest. However,
everything that is developing towards independence must at
some time separate out of the oneness, break apart into *two*.
There arises a duality, a contrast or opposition, in which the one

becomes conscious of itself in the mirror of what it sees. So two always embodies the possibility of development, but it also contains the problems that have to do with freedom and finding oneself: despair, quarrelling, doubt, discord, and lastly the battle between the uniting principle of God and the forces that want to divide and shatter everything, namely forces of evil, the devil as a principle of adversity whose name is associated with division into two. (The word 'devil' and 'diabolic' are etymologically related to the number two.) We also call this the principle of polarization and see it as one of the most important principles of development in the world. This is seen clearly in opposites such as light and darkness, up and down, male and female, stillness and movement, brain and intestines. We can actually hardly imagine a visible world without duality. The difference between the 'clever' head and the apparently stupid 'stomach'[15] — a distinction admittedly relativized today by science — figures in a North German joke: A sailor is sitting in front of several glasses of beer, murmuring to himself, 'Shall I drink another beer, or not? My head says "no" but my stomach says "yes".' The head, being the cleverer of the two, gives way — so he drinks another beer!

A Bulgarian fairytale gives us a picture of the original splitting of the world in two. After God has created the world he rejoices over his creation, but is sad about not having a friend out there with whom he can exchange a word. Suddenly, on his travels, he notices his own shadow, and calls to it, saying: 'Get up, friend!' The shadow gets up in front of him: in appearance like a human being, but in fact the devil. To start with God appreciates this friendship very much and decides to reward him and grant him a wish. True to his nature the devil wishes to have the world divided into two parts. God shall keep heaven and living human beings, but he wants the earth and the dead for himself. The mistrusting devil actually has this confirmed in a written contract, which God later on painfully regrets, for from then on the devil torments human beings with much distress and torment. So the Creator's one-time friend becomes an antagonist.[16]

We see today that whatever kind of polarization there is either in the human being or in the social or political realm works with 'devilish' principles, which is bound to lead to cultural decline or war, if this polarization, necessary as it may be for evolution, is not continually rebalanced again by a third element so that division is healed at a higher level. Here, the three becomes the revelation of a newly developing spirituality, the *trinity*, as we find in many fairytales and myths, whether as the 'third son', the 'third way', or as 'Horus', the 'son' of Isis and Osiris, the new god in Egyptian mythology. This principle becomes externally visible in the pyramids. Threefoldness represents the divine archetypal principles of human nature itself, and thereby becomes the 'open secret' of any cure. It is the 'threefold cosmic seed' of alchemy.

These secrets of numbers which, wrongly applied, can also lead to superstition, also have their humorous aspect. There's a story about an Irishman who was very fond of alcohol and who was born on 3.3.33 in Limerick. While digging peat he struck the ancient and mysterious Ogham Stone which revealed to him that the number 'three' would bring him a great deal of good luck. Closely connected with the mystery of the trinity through his date of birth, he was also born as the third child in the third storey of the third house behind the Church of the Holy Trinity. He discovered whisky when he was 13 years old, women when he was 23, and three days before his thirtieth birthday he inherited from a relative thrice removed, an American tricycle manufacturer, a small fortune consisting of 300 shares of 30 dollars each. He married on the third of the month, was the father of three children, and divided his weekly wages of 30 pounds into three equal piles. By this point, the Irishman MacKibben now had a fair idea that the miraculous stone was right. Three days before his thirty-third birthday MacKibben went into a betting office and placed three times 3000 pounds on the stallion Trinity with the starting number three.

On the third of the third, at 3.30, the race began and Mac-Kibben was already sure of the outcome. But—he lost everything: his savings, his credit worthiness, even, in fact, his life for he suffered a stroke. People can make mistakes, but Ogham Stones never do! The stallion Trinity came in in third place ...[17]

This contrasts with the number four, the 'four mothers', the earth principle: the four elements of earth, water, air and fire, the four temperaments, the four points of the compass, the four seasons, or the salt cube as a representative of the earthly realm with its four corners. In our rectangular houses with flat roofs we see this earthly characteristic especially clearly.

Through the interplay of cosmic-earthly forces in the trinity with fourfoldness, we come to sevenfoldness as the actual 'number of perfection'. The key to the deeper secrets of human being and the world are forged from this. In pictorial terms, the four-cornered house with the triangular roof represents the complete house or the divine body.

Anyone interested in fairytales or familiar with the Apocalypse of John is continually led towards these secrets of number: the sovereign One, the four evangelists, the seven seals and the seven trumpets, the twelve doors or the twelve foundation stones of the new Jerusalem. With these in mind we must recognize that besides neutral, abstract mathematics which we employ in our sums every day and also use to our advantage, there is also another logic of numbers that we justifiably call 'moral logic' or the 'logic of the spirit'.[18]

We will pass over here the secret of the five and six, and return to them elsewhere. However, the number that tells us most about our connection with the cosmos and which we can recognize as the actual number of the macrocosm, is twelve. This is the basic foundation in the visible world whereby our spiritual being, our I, can descend into the physical body and work effectively within it. We find a relationship to this in the twelve senses, the twelve brain nerves, the twelve meridians in Chinese medicine, the twelve apostles of Jesus, King Arthur's twelve knights, the twelve months and the twelve signs of the zodiac.

Taking our guide from bodily nature we can ascend now from oneness through two, three, four, and seven right up to a macrocosmic understanding of twelve, thus fathoming the secret of human nature itself.

We mentioned earlier that everything seeking to come to expression in the sense world must polarize, that is, must split in two, in its physical appearance.

Let us approach the archetypal form of a human being with this thought in mind and take a look at the skeleton. We shall discover two archetypal architectonic principles that have chiefly to do with heavenly and earthly forces: the curved or round forms and the straight, extended ones. We have to learn to 'feel' these two gestures qualitatively. The greater the forces that impinge on a straight line from without the more curved it becomes, and, if it goes as far as becoming a circle, it can enclose and internalize something.

We find the skeleton's most marked curves as a 'cosmic sphere' in the realm of the skull, where our head closes itself off from the outside. In variations (metamorphoses) these enclosures descend from above to below throughout the whole body—'from head to foot': from the head of the joints through the rounded parts of the inner organs and right down to the arches of the feet. At birth and in infancy the round, cosmic forms still predominate. Not until the child grows further into earthly life, and particularly in puberty, does the lengthening growth of the limbs dominate. The human being reaches out towards the earth and 'divides' into fingers and toes. In doing this the open hand corresponds more to the purely earthly form, and when the fingers are drawn inwards the form of the hand is more of a protective, cosmic one, a gesture of inwardness.

We find this polarized principle as *the* archetype of forms particularly at the beginning of our physical existence: as the resting, unifying female ovum and the moving, differentiating male seed. As these two principles unite, the divine is revealed: the third, the human being! The two extreme forms of curved and straight are united in the human being in the middle sphere of the skeleton, as the balance of two extreme principles of rhythmic form in the ribs: higher up they remain closed, and below they open up and extend ever more—twelve pairs altogether!

We find these archetypal cosmic forms of curves, straight lines and rhythmic balance again, in many variations in nature, too. We only have to think of mushrooms with their heads, their rhythmic gills and straight stems, of the dandelion and other plants with round seed heads, straight tap roots, and leaves that grow in rhythms, or also, in the case of human beings, the more rounded female constitution or the more angular male one. From this of course arise interesting medicinal-constitutional aspects regarding health and illness, too, if we think of hydrocephalus, rickets with its protruding convex forms in the skull area, barrel-shaped chest and bent O-legs, but also of the round forms of the mother-to-be in pregnancy.

In round, partly-closed forms, we are really being shown a piece of the cosmos on earth, descending as far as to the arch of the feet, through which we are 'sucked away' a little from the earth. Thinking of it this way, it is easier to understand the anomaly of flat feet and also why animals don't have this arch. In the above-mentioned collection of Bulgarian fairytales about the creation of man, we also read how the arch of the foot came about: An angel sent by God stole back the contract about the living and the dead from the devil, and quickly flew to his creator with it. The devil caught up with him, and caught hold of his foot shortly before he entered paradise, and tore a piece from his sole. He appeared limping before the Lord, who asked what had befallen him. The angel told him what had happened, and how, and in order to save human beings from the devil he had had to lose a piece of his foot. Thereupon God decided to take away a piece of their feet in memory of this deed, and since then our feet have been arched.

The form principles described here are necessary, for instance, for us to close ourselves off from the world in our heads, isolate ourselves, and in doing so develop inwardly. As opposed to the head we find in the limbs the soft parts outside and the bones inside. Thus here we open ourselves to the world through movement. So if someone always 'sits' in his head and scarcely ever moves about, he is bound gradually to develop unworldly thoughts. For this reason Nietzsche once said: '*Sit* as little as possible; don't believe any thoughts that are not born in freedom and whilst moving freely about, in which your muscles also celebrate a festival. Sedentariness is [...] the real sin against the Holy Ghost.'[19]

In the middle, rhythmic realm is to be found, as a third element, a balance between inside and out. Here we are both human being and the world simultaneously, as we can for example experience in our breathing.

The threefoldness we have spoken of is to be found again as a formative principle in miniature, in, say, the three finger joints, the regions of the iris, the outer ear or the inner organs. It is the physical reflection of what was first described by Rudolf Steiner in 1917 as the 'threefold (not tripartite) organism': in the upper part of the human being is the nervous system (head), in the middle realm the rhythmic system (chest with the heart and lungs) and in lower realm lies the metabolic system, inward-oriented in its opening to and internalizing of the outer world, but outward-oriented in its foundation for the movements of the limbs. These three systems are of course active throughout the organism, but have their specific centres in the three realms mentioned. The whole body with its three systems can be found again in a striking way especially in the human head: in the enclosed realm of the upper skull, the actual area of thinking; in the middle region of the sense organs, through which we grasp the world—seven 'apertures' through which the world flows into us: two for the eyes, two for the nose, two for the ears and one for the mouth; and in the lower jaw and mouth area we find the element of movement, the beginning of metabolic activity in the chewing and mashing jaw as the entrance for food. So we can rediscover our whole organism again in the head, and the head in turn in the whole human being. Recognizing this, we discover interesting relationships between different areas of function.

So the 'face' of our whole bodily frame has a 'forehead'—the head; has 'eyes'—the arms, with which we reach out into the world; has a 'nose'—the chest that expands into the world as it breathes; and has a 'mouth'—metabolism or the 'desire sphere' with its inner connection to both the abdomen and also sexuality. The lower jaw with its joints corresponds to the legs. If we go into detail we also find in the oral cavity a threefold principle: in the upper jaw the round head forces, in the teeth solidified rhythm, and in the tongue — which in some abnormal conditions, for instance Down's syndrome, becomes disproportionally large — the muscle-limb system.

In the sketches relating to this, Rudolf Steiner noted of the head, chest area and limb system that the head is a 'passenger' carried around in a 'sedan-chair' as though it comes from 'beyond the earth'.

The chest that breathes and the abdomen that receives our nourishment are like a child enjoying its suckling, letting itself be looked after by someone outside it, by a wet-nurse if you like. The limbs are busy workers, however, harnessed to an 'earthly yoke', and like slaves have to carry the entire 'middle and upper realm' through life. Human beings are rejuvenated by cosmic energy. Only the earth makes them old. The head, as the representative of the cosmos, should not actually grow old, for it, in Steiner's words, remains forever a 'Peter Pan'.

It is only a small step from here to start understanding that these three principles of education and *function* are also the bodily basis of our three soul forces: thinking in the head, feeling in the chest and will through the limbs.

These three *basic* principles represent one of the most important diagnostic keys and therapy guidelines in anthroposophically-extended medicine and psychotherapy. It took Rudolf Steiner decades of research into the human being, nature and the social organism to develop the threefold social order and to work out its specific details. In olden times already of course we find references to threefold qualities in nature and in the human kingdom. In alchemy three elements represent the forces of the head, the rhythmic system and the metabolic system: 'Salt' stands for the head and for thinking, 'Mercury' the divine messenger between heaven and earth, for the rhythmic element as the basis of feeling, and 'Sulphur' or 'Phosphor', the bearer of light and warmth, that brings everything that is earthly into movement, for dissolving and warming metabolic activities.

In Christian symbolism we find these three functional areas represented by animals that stand for thinking, feeling and will: eagle, lion and bull, that unite with a fourth, the human being, the 'waterman', coming to completion in the human face. This 'fourfold animal', which represents the whole cosmic human being, is also brought into a connection with the four evangelists: the eagle with John, the lion with Mark, the bull with Luke, and the human being with Matthew. This relationship which we can trace right into the language of each of the Gospels, is very important for the Christian culture of the West. To illustrate this, I will quote the beginning of each Gospel below.[20]

John, the evangelist of the Logos, of thinking (eagle): 'In the beginning was the Word, and the Word was with God, and the Word was God.'

Mark, the evangelist of energetic courage for knowledge, (lion); 'The new Word of the angelic realms dawns through Jesus Christ.'[*]

Luke, the doctor, (bull) who has a loving interest in human descent, that is, in ancestors, speaks in an incisive way about the pregnancy and birth of John and Jesus.

Matthew, (waterman), envisioning the incarnation of a divine being: 'This is the book of the human incarnation of Jesus Christ...'

In other cultures, too, eagle, lion and cows or bulls are particularly venerated.

We have now arrived at fourfoldness, as represented in bodily nature since olden times particularly by the four elements. Earth, water, air and fire, are the earthly expression of the human being's four realms of existence: the earth for the physical, water for living things, air for the soul and fire for the spirit. The pattern of these four elements in the human organism is again dependent on our soul-spiritual activities. For instance *warmth* lives in enthusiasm, *coolness* in abstract thoughts or in coldness of feeling, *rigidity* in fear, *flow* in the motions and stirrings of the soul, and *air* in the 'happy-go-lucky' temperament. When these qualities become chaotic, or when one predominates over another, this leads to illnesses. But our inner elements also remain in a close

---

[*] The beginnings of each Gospel given in German differ from the King James versions, and the translator has therefore translated directly from the German here.

relationship with outer ones, as we shall see in the case of the four 'meteorological' organs of the lungs, liver, kidney/bladder and heart. This is clear when we have a cold: the four elements within us separate out so that, 'poisoned' by external cold, the warmth organism reacts with a temperature, the air organism with a cough, the water organism with a head cold or excessive mucus formation, and the earth organism in the worst cases with the destruction of the mucous membrane. This is why cures using the four elements have been an integral part of medicine in all cultures for millennia.

Before we now proceed to the inner realm of life — the seven chief organs of a human being that, as well as being connected with the seven life movements, also express themselves among other things in the seven hormone glands or chakras, through which the soul element can combine with life forces — let us first consider a great macrocosmic secret of our bodily structure: the zodiac constituting us.

The human being as constituted by the zodiac

In accordance with the threefold form of the human organism, there are three basic functions connected respectively with the development of the head, the rhythmic system and the meta-bolic-limb system:

- Self-enclosure or internal assimilation of the world
- Interior expansion or the outspreading of the organs
- Self-surrender to the earth through movement

We can see these functions as synonymous with the soul forces of thinking, feeling and will referred to above. As we see from the illustration dating from the seventeenth century (see frontis-piece), the bodily anatomy of the human being is the con-sequence of the effects of the twelve forces that have their cosmic origin in the twelve signs of the zodiac. Even when considering tangible anatomical regions, we still have to recognize the creative impulses lying behind them as gestures that shape the entire body. For each form that has come to rest arises out of living movement that culminated in anatomy. So we can distinguish four functional gestures in each of the three regions — the upper head system, the middle rhythmic system and the lower 'earthly' system.

Thus the forces of Aries, Taurus, Gemini and Cancer are necessary for arriving at the kind of interiority shut off from the world that is a faithful image of the whole universe. This gives the head its distinctive quality, with all its senses, the brain, the skull and surrounding skin as a barrier against the world. In the head the whole world, the universe, in fact, can arise once more in thought. In this respect it becomes understandable that in our 'upper realm' the zodiac is reflected in us in purely dynamic gestures.

Aries, springing forth from the universe and making itself independent, yet gazing back into the universe, takes the whole universe individually into itself. We recognize Aries in the upper head region, the forehead, thus in thinking. This is why its horns, the 'antennae of the cosmos', are curved inwards.

Taurus, situated in the larynx and the neck region, does not look backward as Aries does, but into the whole surroundings of the world, and brings fertile life and movement into our thoughts that then come to expression by way of the larynx as living speech. 'In the beginning was the Word ...'

Gemini, situated on the arms in the human zodiac, makes us inward, in that we develop a left and a right and can encompass not only the outer world but also our own self. This is the reason also why we have two different sides to our face, which becomes evident if, using a photograph of one half of your face, you reflect and project it on the other. Already when we simply cross our arms or put one leg over the other, we close ourselves off from the world, become more inward. Or think of how we perceive the world with our eyes: when we stare at a certain point, the left and right axis of the eyes cross. When we pray we put our hands together, to gather ourselves inwardly. All awareness of our ego, our inwardness, depends basically on this gathering of ourselves together.

Our inner being is enclosed by the skin as a limiting surface (or also the skull): here Cancer, symbolically situated in the upper breast region, acts in the region of the ribs.

From these four cosmic gestures is formed what we call *self-enclosing head activity*, though its forces also work throughout the organism and right into the ribs.

It is this that makes possible an interiority vis-à-vis the world that can be individually filled by the world of feeling. Leo, Virgo, Libra and Scorpio can now carry their work further.

The functional gesture of Leo in bodily development is situated at the all-governing centre, that is, in the heart and the circulation that pulsates through the whole organism. Thus the heart embodies the archetype of feeling, inwardness and rhythm.

Everything that arises within, however, moves towards its completion, becomes a seed for something new, so it must at some time wither, to become nourishment like corn. This is Virgo with the withering ear of corn in her hand, who resides in the lower small intestine and kidney/bladder region where once-living nourishment goes towards its end.

By means of the substance we take in we become more earthly, we integrate ourselves again into the mineral world, and must find a balance between ourselves and the forces of nature, for instance gravity. This is why Libra is situated in the region of the hips and the lower section of the intestine.

But what would we be as human beings if we were not daily imbibing from the world things outside us such as light, air and nourishment, that is, influences beyond the human world that also threaten our inner nature and which can turn into something foreign in the form of a 'poison', against which we would have to defend ourselves with all our strength. Scorpio with its poisonous sting sits deep in our lower bodily metabolic-sexual region, which – if we think for instance of sexually-transmitted diseases or Aids – can be a point of entry for destructive forces. A foreign cosmic world is pushing its way deeply into our organism here which is why, on the other hand, this is also the place where in the case of women fertilization from without can occur and planetary forces attend to the development of the human body.

But generally speaking Scorpio has something to do with toxicity and detoxification, and is thus a seam of closest contact between inwardness and the external world. Inwardness can only arise if it defends itself against an external influence and in doing so develops its own forces.

Only now can we plunge our limbs, that is, our legs, into the earthly world, and in doing so enter into gravity with thighs, knees, lower legs and feet. These represent the most diverse kinds of purely earthly activity. Just as we are enlarged and extended in the head region, as earthly human beings we are restricted in our legs to specific earthly activities.

From spiritual-scientific, anthroposophical research we know that the fourfoldness of the legs has to do with the four archetypal professions of humanity, one could also say with a gradual process of growing into the world of human culture. Thus the thighs formed by Sagittarius, the knees by Capricorn, the lower legs by Aquarius and the feet by Pisces reveal the archetypal professions of hunter, animal husbander, farmer and trader, thus passing from untamed nature through to the cultural world that human beings have created.

The hunter, who lives out of the force of his thighs, and who hunts wild animals, does not yet find a world civilized by human beings. He is the marksman who emerges from an animal body. Only when he tames the animals and refines them, bows down to them and engages his knees, does a more settled agriculture commence. The original image of Capricorn thus had a fish tail as a symbol of the fact that man has changed wild animals by breeding them.

Now it becomes possible to develop agriculture and a settled way of life, and human beings walk through their fields with containers of water and stand up to their lower thighs in the liquid element to make the fields fruitful.

But what do they do with the harvested fruits? They have to dispose of them, sell them. In earlier times the feet of Pisces were portrayed as a pair of ships alongside one another, as a picture of someone who crosses the ocean to trade. It is with our feet that we move about in the world and go towards other people. In life we have to stand upright in support of something, get onto our own feet and 'walk' our path of destiny. With the soles of our feet we are connected most fully with the earth.

We can now really ask ourselves whether this human being composed of the signs of the zodiac in any way enlarges our knowledge today or whether it is just a beautiful but very outmoded image. We might think in this connection of the composer Gustav Mahler, who once said: 'Tradition is not the worship of ashes, but the passing on of fire.'

In our view it certainly is worth occupying ourselves a bit more with this imaginative picture. It is not so rare, in fact, to discover that the astrological star signs do have a relationship to the organ areas corresponding to them, whether as a strength or an organic weakness.

Sometimes the organ areas belonging to someone's astrological signs are especially well developed, for instance the forehead in the case of Aries, the neck region in the case of Taurus (a bull neck) the breast region in Leo, the thigh in Sagittarius and the calf in Aquarius.

During my work as a doctor I have often seen weaknesses and even illnesses in the organ regions connected with a patient's star sign. For instance, I have found cases of severe hip disorder in people born in the sign of Libra—one person with a dislocation of the hip joint and another with hip joint necrosis; and the grandmother of the first patient, who was born in Libra, suffered from the same hip complaint.

Another patient with the star sign Taurus had a typical bull neck, spoke in a well-formed but loud voice, but repeatedly suffered from a severe case of illness of the larynx and changes in the thyroid gland. I also remember someone born under Gemini who always had problems in the upper shoulder, and another patient born under Capricorn who reacted to the massive hindrances she encountered with severe psychosomatic knee problems. None of this inevitably occurs, but it can do so; and this may help us develop a deeper understanding for some illnesses.

The human zodiac is especially valuable if we consider the

polar connection or opposition between organ areas, which anatomy regards as being disconnected from one another. Then we can understand better some of the connections that are known to us from experience.

Here are a few examples of this:

The polar opposite star sign to Taurus (as a life sign) is Scorpio, that is, the region of the larynx and the sexual system. We experience this connection no later than puberty, when the voice begins to break, and sexual maturity is starting. Likewise suppressed erotic urges can manifest in their 'opposite' region as loss of voice, a chronic urge to clear one's throat, or massive swelling of the thyroid gland. I believe that everyone knows from experience how strongly the voice and erotic feeling belong together.

Then we have the polarity between Leo and Aquarius, that is, heart/circulation and the calves. Heart weaknesses manifest as oedema in the calves, and circulatory disorders lead to calf cramps at night; but equally, there is no better remedy for heart and circulation weakness than calf massage to stimulate or calm a person down. Since the lower arms correspond to the calves, cold lemon compresses on pulse surfaces and lower arms are very helpful in the case of circulatory disturbances.

As to the polarity between Virgo and Pisces, relating to the kidney/bladder region and the feet, we know the inner connection between wet and cold feet and subsequent abdominal colds affecting the bladder or kidney region. I know of no better treatment for these conditions than hot footbaths! But a treatment of the feet that benefits the abdomen also works curatively on the next 'level up', the head.

Nietzsche was one of the great thinkers who never ceased wondering at 'the intelligence of the body'. He once wrote that this whole phenomenon 'body' is as superior to our conscious thinking, feeling and will as algebra is to our multiplication tables.

He was also clearly conscious of the vast difference there is between the body that we feel, see, fear and wonder at, and the 'body' that anatomy teaches us about. According to his view it is much more astonishing than human consciousness. We cannot sufficiently admire how this body became possible: how living forces unite in such a tremendous way, each dependent and subject and yet in a sense acting and commanding by their own volition, and how they live as a harmonious whole, growing and persisting in this wisdom over many decades—yet apparently without our conscious involvement!

According to Nietzsche, therefore, everything that originates from purely intellectual self-reflection is unproductive. For without the sure guidance of the body he did not believe anything good would emerge from human enquiry. In his view, the philosophical knowledge of the future must have something 'bodily' about it, and pass from the abstract concept of 'true' or 'untrue' to 'healthy' or 'unhealthy'.

We shall now turn to the world of the inner organs, whose sevenfold membering corresponds to the forces of the planets— four of which, as we have already mentioned, are especially 'open' to the outer meteorological processes of earth, water, air and warmth. Just as our spiritual being, our I, needs twelvefoldness to gain a full purchase on the earthly realm, so our soul requires the sevenfold levels of the life forces to be able to reside in our bodily nature. Thus we must first of all ask ourselves: What do we really understand by an 'organ'?

The Planetary Order in the Organs

If we lose heaven we lose ourself.[21]

Rudolf Steiner

Can we come to a real, comprehensive understanding of the organs if we envisage the various organs as organized piles of cells, like bricks lying unconnected one beside the other — or like embalmed members of a family who in life never had a closer social contact with one another, thus forging a shared family destiny? Just as each single member of a family has their own history, character, preferences, and closer or more distant connections within a social context, the organs, beside their anatomical, physiological aspect, also have relationships and functions throughout the organism, even also reflected in the human soul. Therefore in traditional Chinese medicine one speaks of 'organ families'.

Let us take as an example a 'heavenly organ' like the moon, our earth satellite, which we will here for a moment consider in terms analogous to one of our inner organs.

As a physically perceptible 'object' we can determine its distance and size and also determine certain rhythmically arising changes. Yet we actually never see the whole moon, but always only part of it, as it hides its far side from us. But does the moon also have a living functional meaning that goes beyond its physical position? We can experience this on the earth if we study its influence on the weather, the tides, groundwater behaviour, a woman's menstruation, the growth of plants, and the activity of microscopic creatures in the ground at full or new moon. In some regions of Europe it is still the custom today among farmers only to cut down wood at certain phases of the moon because of the lunar effect on moisture content. Anyone who has worked in midwifery also knows that the period of full moon has an influence here.

As well as its effects at the 'life plane', does the moon also exert a 'soul effect' upon us? It is well known that some people can't fall asleep at full moon, have bad dreams or even sleepwalk. It is possible to experience its influence on our desires, imagination, and *mood* (*Laune*, the German for mood, derives from the Latin 'Luna'). And women sometimes get into moods before or during menstruation. This is also why a crazy person is called a 'lunatic'. In the USA a psychiatrist compared case histories in psychiatric clinics over a period of more than 30 years with the phases of the moon, and established the fact that at new moon and full moon clinic admissions always rose steeply.

In the case of each planet or satellite we can assume a spiritual effect if we study its influence in harmonious interplay with other planets. Thousands of years ago people referred to such influences by the names of gods such as Saturn, Jupiter or Mercury, these being thought to affect both heaven and earth. Let us look more closely at this superordinate aspect that works right into the order of the human organs and even into plants and metals and our soul structure, in order to gain better under-standing of the individual organs and soul functions in health and illness. For instance in the USA not long ago, based on Jungian psychology, people sought to revive a cosmological view of the soul, as propounded by Marsilio Ficino, a Renaissance scholar (1489), and tried to discern certain psychological pro-cesses in terms of planetary influences. Such an approach can help us see why, in olden times, the soul was called a 'star body' — a 'sidereal' or 'astral' body.[22]

There have always been attempts in humanity to apply to the organs or their processes of illness the fourfold approach at the different levels of body, life, soul and spirit. Thus we find in the work by Paracelsus, *Volumen Paramirum,* the account of four doctors standing around someone who has died of cholera, where each of them describes from his own point of view the cause of death: The *material* cause of illness — as today we would attribute the illness to an infection — the *constitutional* state of the patient, connected with a person's life forces or immune resi-lience, *psychological* impairments, for instance grief, worry, depression, and the level of the *individual ego,* for instance in its structures of fear, something which is investigated today in the field of psychoneuroimmunology. For Paracelsus, all this culminates in a superordinate fifth level, the 'quintessence', in the laws of individual destiny which are also called karma.

We can study the individual human organs in these varying aspects, to obtain as full as possible a picture of their function within the whole organism, and to find therapeutic measures that accord with these.

The physical, sensory expression of an organ, its size, form, colour or position already says a great deal about its nature. But we can learn much more from its physiological and biochemical life activity, its rhythm — the 'organ clock' as Chinese doctors call it — or even from the influence on mind and soul we can intuit when we feel 'liverish' or are so angry that 'we spit bile' or when we think someone is a 'person of my own kidney'. Why should these soul forces found in the organs not fit together like individual stones in a mosaic to constitute the whole of our soul life, with the brain just the conscious projection screen for nuances of feeling arising from the organs? For thousands of years Chinese medicine has associated the heart with joy and the ability to love, the lungs with sadness, the kidneys with fear and the liver and gall with anger, and Nietzsche, who is well known for his radical sayings, coined the phrase 'All our prejudices originate in our innards ...' (op. cit.)

We can try imagining our consciousness as a bright, mirroring pond, into which flow from our organs the hues that belong to each so that we can develop an individual life of feeling. But alas if one colour grows too dominant and the whole soul lake turns dark blue (we think of sadness as blue, don't we?) — or a poisonous green.

We mentioned elsewhere that a person's life forces are sevenfold, allowing the soul to connect with a corresponding bodily foundation. This is why in the case of so-called psychosomatic symptoms, seven as the 'number of perfection' plays such a great role in the healthy collaboration of life forces and psyche. It is always the necessary conclusion of a stage of development, and thus the precondition for a new developmental process. This is also seen in music with its seven tones as the preparation for the next stage, the octave. This is why there are seven days of Creation with the day of rest, Sunday, as a preparation for the following week.* We also see this in the cycle of seven years in the biographical development of a human being or in the mirroring of the events of life. Even folk wisdom hints, jokingly, at sevenfoldness in the saying that a cold treated by a doctor takes seven days and, without this, a week. The proverbial 'dreaded seventh year' of a relationship also belongs in a way to this. In this connection it is also interesting that in the USA, in comparisons of 30,000 people, the ideal sleeping time for health and soul well-being was found to be around seven hours, though depending on age and physical constitution this length can naturally vary. This fact has been known for a long time also in anthroposophically oriented medicine. Seven is significant as the number of perfection and preparation for a new beginning — also when doubled, trebled or halved (3.5) — even in terms of social phenomena, as we can see in the development of institutions. Economic research has likewise confirmed that seven lean years usually follow seven bountiful ones. Many myths and fairytales point, either directly or indirectly, to the secret of sevenfoldness in our life and soul forces. For instance the seven dwarfs in the story of 'Snow White' represent healing forces that embody earth wisdom relating to the planets. In some versions of the tale in German, the dwarfs are given names that each hint at a different planetary influence.[23]

Below, in trying to describe the inner relationship of the organs to the seven planets we must be conscious that the relationship goes far beyond a physiological or anatomical description. An 'organ' is the visible culmination of a comprehensive process working in the bodily and soul functions of the whole organism, so that it would be more accurate to speak of 'organ processes', or 'planetary processes' that come to a physical conclusion in the seven 'metal processes' of the earth. For example, Goethe writes

---

*Paracelsus, in fact, speaks of four days for the work of the hands, and three days for the work of the heart.

in his *West-East Divan*: 'The planets, all seven, have opened the metal gates ...' In earlier times – and even today still in certain older sayings or traditions – the metals were connected with the planets: silver with the moon, gold with the sun, quicksilver with Mercury and lead with Saturn. Lead poisoning was called *'morbus saturnicus'*. In modern colloquialisms we still characterize certain qualities in terms of metals: 'heavy as lead', an 'iron will', a 'golden boy', 'nimble as quicksilver', 'silver-tongued' ...

To begin with it is not so easy to understand that all our organ processes, like the metals, embody a dual, polar activity, with a necessary distinction between outside and inside. We shall see this later on when describing the single organ processes in greater detail. We therefore speak of 'dual' planetary processes or organ activities and can compare this with burning wood that falls into ashes downwards but develops a blazing heat upwards. A plant, too, conceals in its roots and blossom at the same time both an earthly occurrence and a cosmic one. This thought is important for correcting a one-dimensional and largely outward view of the organ activities. We can therefore ask for example: What is the inner action of the kidneys, that outwardly eliminate water? Or the liver, or the brain? Or what happens in the organism when I give lead (plumbum) as a medication in a low homeopathic dose, or a higher one? Doesn't this actually achieve opposite effects, although based on the same substance?

Here we can recognize that lifelong study and a great deal of experience is needed to penetrate into complex organ processes and the corresponding metal processes and handle them correctly. Rudolf Steiner once described a human being as 'sevenfold metal'.

However great the merit of homeopathy in its empirical testing on human subjects of metals and their chemical combinations to produce 'medicine pictures', we shall see that only spiritual-scientific insight into the organs, and their relation to planetary forces, provides a deeper foundation for therapy. This is no slight on homeopathy, which the author himself greatly esteems.

The spleen as the Saturn organ

The first universal functional gesture for our individual earthly embodiment is separation and self-enclosing from the outer world, so as to be able to develop a personal inner life. However, everything that 'falls' from the cosmos to the earth is subject to a temporal limitation, that is, death, which is necessary in order to bring the human being's spiritual part back to his true home. So, basically, incarnating means that the spirit 'dies' into solid earthly form, and makes for itself in the skeleton, the symbolic picture of death, its solid earthly image.

Through our purely physical sense organs, also, our living impressions die into dead ideas, which we then preserve in the 'mummy chamber' of our memory. This self-enclosure and the possibility of becoming a self in the course of time was called 'Saturn' or 'Chronos' by earlier people, taking effect in a successive 'chronology'. This is why the physical planet Saturn is situated at the border of the cosmos between, as it were, spiritualization (death) and embodiment (life) and is surrounded by a ring that encloses it and enables it to have an individualized physical and psychological activity of its own. In Greek mythology Saturn-Chronos engenders children whom he proceeds to eat. We find these very functions in the spleen, the true Saturn organ. Towards the outside it protects our inwardness by embodying a 'bulwark' against the invasion of foreign bodies, that is, it is one of the most important organs of immunity; within, by contrast, it devours its 'children', the blood – which at one time in the embryonic period it was engaged in producing – in that after about 80 to 100 days it destroys the old blood corpuscles and conducts them to the liver, which turns them into bile pigments (bilirubin). Through this cleansing process the blood can recreate itself anew and revivify. Can we imagine a greater metamorphosis than occurs in the spleen, where once sweet, red, living blood becomes, with the help of the liver, a bitter viscous greenish fluid, bile, that breaks down food? From this it is evident how closely spleen, liver and bile formation work together. It has long been known in spiritual-scientific medicine that the spleen also has rhythmically regulating functions in the upper abdomen in that it has a balancing effect both in the blood's storing functions and in its relation to the liver portal vein system, so that when intake of nourishment is irregular, the inner rhythms that we need for organic independence are assured. The spleen—which in olden times was seen as the organ of melancholy, the temperament of the depths (Greek: *melanos cholos* = black bile)—is the most spiritual of organs and can be removed with relatively little ill effects. It continues to present medicine with enigmas, as the following episode shows. An American professor of pathology, Milton Winterwitz, once asked an unprepared student to describe the tasks of the spleen. Winterwitz was, by the way, feared by students for his severe, sarcastic manner. The astonished student said in answer: 'Oh—I know them all, yes I do, Dr Winternitz. In fact, I was reading about the spleen just last night and I knew all there is to know about it, but I'm drawing a blank. I've forgotten everything.'

Winterwitz is said to have answered with a scornful grin: 'Why, what a great loss to medical science. Just think—the only person in history ever to know what the spleen does, and he's forgotten it!'[24]

From an organic and psychological point of view the spleen, then, stands for organic and psychological independence, self-demarcation and individualization. As an immunologist once put it so aptly, from a bodily point of view the immune system helps us to divide ourselves as an 'I' from the 'not I', that is, from our environment. In this respect it can be stimulated in cases of a weakness of the immune system and digestive weaknesses, but also when the most unconscious, spiritual part in the organism, the instincts, fail. It is interesting, in this case, that the healthy Saturn gesture we have described, when it deviates upwards or downwards, leads, also in psychological terms, to a failing or exaggerated self-demarcation. If a person cannot fend off outer impressions, and is at risk of being inundated by them, a psychological or also a bodily lack of self-protection arises, for instance in irritations caused by the weather or other outer influences. Conversely, people can become so self-willed that they shut themselves off completely from the world, as we know in what used to be called 'spleen' — meaning bad temper or spite. Healthy spleen activity accordingly means an organic and psychological self-enclosure, and at the same time an inner 'assimilation' of what has been taken in, which is incorporated into one's own being.

Now it will also become clear why lead was assigned to Saturn and why it has a therapeutic significance as a homeopathic medicine. If it works too strongly as a poison, it kills people by rendering them too 'ossified', that is, it makes them sclerotic, too cold — psychologically as well as physically. If its effect is too weak, this leads to dissolution and softness inadequate for developing bone formation necessary for earthly life. We see that lead, with its 'dual planetary process' properly applied, can be used against both sclerosis as well as rickets. As we see in the case of a lead shirt against x rays, it protects us with its 'ring' against too strong an external psychological or physical influence. Therefore it is, among other things, also helpful against allergies. In my practice I once experienced a striking example of this. A patient had not only suffered brain concussion in an accident but also suffered from severe insomnia, because he kept reliving overpowering auditory impressions of cracking and splintering. An injection of lead D 20 put a quick end to this problem, killing off these over-active impressions.

In the head region the lead process works at its strongest at the back of the head, where we divide ourselves from the world most of all, and where the bones are softest in the case of rickets. This rays in in the area of the fontanel, where monks have their tonsure and Jews their kippah, and where a man's hair often starts to fall out. If we touch this place then, in comparison to the rest of the head, we feel that this is where warmth streams in. From this point of view it becomes clear that, as far as warmth conditions of the whole organism are concerned, especially in the case of small children but also for people who do intensive mental work, this area should be especially protected.

The Saturn process is interesting also from the point of view of its biographical effect. This planet takes 30 years to orbit the sun.

We can often see in human development that a crisis occurs around the age of 30: what is old comes to an end, so that something new and personal can form. This often happens psychologically in that something instinctively familiar either dies away by itself or is unconsciously resisted. A person has then, as it were, to go through an inner death in order to be born anew in the spirit.

So Saturn stands for healthy development at a time when one should neither get old too prematurely nor wallow too long in youthfulness.

In Chinese medicine, and in anthroposophic medicine too, the spleen has a major functional significance in terms of energy, despite the fact that, as a purely physical organ, it is more or less dispensable. Apart from a certain immunity weakness that mostly passes quickly, there are hardly any serious complications to be expected after its removal. So we can ask why it is different in this respect from hormone glands, where each molecule of physical substance is critical? Might there be organs within us in which a spiritual energy is not yet entirely bound to the physical?

In Chinese medicine the spleen is responsible for the transformation and transportation of food into the blood, and can even absorb cosmic vibrations, i.e., the 'thoughts of the cosmos'. As such, it can transmit the life energy in our food to the blood, and through its rhythmic function individually harmonize the inner motions of our metabolic organs; its effect on the blood even extends as far as the rhythmic functions of the uterus by regulating the rhythm and length of time of menstruation. It is also worth noticing that in Asiatic medicine the spleen (and the pancreas too) is connected with the 'sweet' taste. After a meal there sometimes arises a need for something sweet that has a strengthening effect on the digestion and, in the anthroposophic view, strengthens the ego activity in digestion. The period of prime energy for the spleen and the pancreas is in the morning between 9 and 11 am, following which comes a small dip in energy – which as we well know is overcome in the UK with 'elevenses'. In the Chinese 'organ clock', from then until 1 pm begins the cardio-vascular energy period. Around this time it is a good thing, in cases of low blood pressure or a decrease in efficiency, to drink a hot salty broth. Salt is connected with the head, helps our concentration and certain forms of headache, warms us up and raises blood pressure.

Conscious chewing and above all regular intake of food relieves and supports metabolic spleen activity in its relation to the blood. This makes it an important border organ between the inner and outer world! For the border between inside and outside is important for our identity, as we saw in the protective 'ring' of our immune system. Through our breathing rhythm we still live of course in a pronounced world rhythm, which is why the breath is known as the 'great healer'. For, at an average of 18 breaths a minute × 60 an hour, we take around 25,920 breaths a day, a number identical with the years that the sun requires to pass through the whole zodiac (called the 'Platonic year').

Our often random — that is arrhythmic — intake of food means that we are at risk, especially in this era of nervous haste, to lapse increasingly from a harmonious connection with the world (a 'neglect of biorhythms').

Thus the spleen mediates between randomness and constancy.

> The spleen is ... largely an unconscious organ, and it reacts to an extraordinarily great degree to the rhythm of the intake of human food. People who are eating all the time, call up in themselves quite a different spleen activity than people who also allow in-between times. We can notice this especially in fidgety spleen activity in children when they are perpetually nibbling; a strongly fidgety spleen activity develops.[25]

Thus the spleen mediates between a health-inducing breathing rhythm and an unhealthy arrhythmical one of food assimilation. We know the results of this arrhythmical behaviour on the metabolism: burping, flatulence, various food intolerances, heartburn etc. These are not only attributable to the stomach!

The unconscious spirituality of the spleen, however, regulates an important ability that we absolutely need for keeping our body healthy: the food instinct! Instincts are 'unconscious will conditions'. Through the weakening of splenic energy the healthy instincts are gradually corrupted—an important theme especially in the way children are educated today.

In order to strengthen or make the spleen healthy, anthroposophic medicine recommends using the corresponding planetary metal, lead, in the form of a lead ointment (plumbum). This is gently rubbed onto the spleen area, complemented by a kind of inner 'spleen massage': smaller meals, at more regular intervals, with a rest after a meal. This is the way to treat various weaknesses of the metabolism. In some countries, like Switzerland, it is still traditional to have five meals a day. Breakfast, a small meal around 9 am; lunch; an afternoon meal around 4 pm; and then supper. In the first of Rudolf Steiner's medical courses, he says in this regard:

> You see, in our frenetic age, where people are actually always—at least a lot of them—busily caught up in an activity that wears them out, the function of the spleen is greatly affected by this activity whilst a person is engaged in it. Unlike certain animals, that keep themselves healthy by lying down, not allowing the digestion to be disturbed by outer activity—they really spare their spleen much labour—a human being rushes about in a nervous, frantic way, with no regard for the spleen. This is why, throughout the civilized world, spleen activity has gradually become very abnormal; and it becomes very important to relieve spleen function through the measures that I have been describing.[26]

If the soul aspect of spleen function is excessively active, then as we already saw, excessive self-encapsulation occurs in the form of cranky or whimsical behaviour, but also bad moods, grumpiness or depression. A person will become too hardened, isolated, that is too saturnine. The counterpart to this, the polar aspect of Saturn as it were, is to ascend from weighty corporeality through joy and interest, thus opening oneself to the world. As the most spiritual organ in the human being the spleen inspires a quality that connects a person, as a spiritually creative being, with the world in the most intense and inward warmth: enthusiasm lifts a person out of the 'heavy mass' of the body.

Friedrich Schiller, who had a great many saturnine features, right into his bodily constitution and habits, exemplifies what a health-bringing effect enthusiasm can have even on sick organs scarcely capable of functioning any more:

> After Schiller's death 9 May 1805, a post-mortem found that the lungs were gangrenous, mushy and quite 'disorganized', his heart had no muscle substance, the gall bladder and the spleen were unnaturally enlarged, the substance of the kidneys was in a state of dissolution and totally deformed.

Dr Hauschke, the private doctor of the Duke of Weimar, added this succinct phrase to the findings of the post-mortem: 'Under these conditions we must be amazed that the poor man could live so long.'

Schiller himself said that 'It is the spirit which builds the body.' Obviously that was true in his case. His creative enthusiasm kept him alive beyond the date when he should have died. Heinrich Voss, who was with Schiller when he died, wrote: 'The length of his life is attributable only to his infinite spirit.'

The findings of the post-mortem can demonstrate one definition of Schiller's idealism:

> Idealism is when the power of enthusiasm allows us to live longer than the body would otherwise allow. It is the triumph of a clear, enlightened will.[27]

The liver as the Jupiter organ

If I were informed tomorrow that I was in direct communication with my liver, and could now take over, I would become deeply depressed. I'd sooner be told, forty thousand feet over Denver, that the 747 jet in which I had a coach seat was now mine to operate as I pleased; at least I would have the hope of bailing out, if I could find a parachute and discover quickly how to open a door. Nothing would save me and my liver, if I were in charge. For I am, to face the facts squarely, considerably less intelligent than my liver.'[28]

The pain of the liver is tiredness. Folk wisdom

In Greek mythology, Zeus/Jupiter (or Jove) becomes the successor to Saturn/Chronos who devours his children, and who was deceived by being given a stone to devour instead of a child (here again we see the relationship of Saturn to the solid-mineral element). As the highest god of a new generation Zeus stands for wisdom, circumspection, law, form and energetic life, as we would expect of a 'jovial' person. He is as responsible for regulating processes of life and renewal in the world as the liver, the 'general' in our organism. Jupiter's planetary activity consists in bringing form and structure from above into the organism. This is why, in the head region, in the curve of the forehead — which, when pronounced, we sometimes call a 'Jupiter forehead' — we find the outer expression of intelligence and wisdom. The pituitary too, with its regulating influence on most of the subordinate hormone glands, is associated with this Jupiter activity. As we have already mentioned elsewhere, we find in the curves of the various organs and the skeleton the expression of a cosmic 'sculpting' activity originating in our upper realm. If the organism becomes too hard, relatively too sclerotic, or if it remains too soft, then it becomes impossible to fully sculpt and shape the physical body. So the Jupiter process preserves structures in a balance between too fluid and too dry. The corresponding metal is tin (stannum). In its chemical reactions it maintains an ideal balance between firm and liquid, and for this reason was used for the production of tins for the food industry. (Tin of course refers both to the metal and the container to put things in.) A lot of tin is found in peninsulas near the sea, that is, where water and land meet. In the organism the tongue, as the organ of taste, is a muscle rich in tin, which, like a 'peninsula', is washed by liquid, saliva. This is a phenomenological indication that the taste experience is connected with the liver process, for we can taste only when we dissolve substance in saliva. So we can by all means say that the liver behaves in its many functions as a sensitive 'organ of taste', tasting the portion of the outer world we take in as food so that it can then decide, as a 'general' for the whole organism, what should happen with the substances so that, where necessary, it can eliminate alien matter from the organism. Through its connection with one of the biggest veins in the organism, the portal vein, we can also regard the liver as a venous organ ruling all the veins. The Chinese meridian runs along the inside of the upper thigh via the great vein (vena saphena) in which thrombosis and inflammations often occur, which is why, in the case of venous disorders, liver-supportive treatment with medicines such as Milkthistle seeds appears to be indicated.

It is interesting that traditional Chinese medicine assigns to this 'general', the liver, both the season of spring with its burgeoning life forces, and the element of 'wood', whose symbol is a tree. Taking up this picture we can arrive at a more comprehensive understanding of liver activity. A healthy tree takes root firmly in the ground, out of which it obtains its water, and is surrounded by bark, which, at the outset of growth, is neither too damp nor too dry. In relation to the seasons a tree is extremely adaptable, supple and stable. This inner, organic 'sappiness' and constancy is imparted to us by the liver. It earths us as we digest the food we take in. This cosmic organ of water and life is, by virtue of our earthly desire for enjoyment, chained to the earth like the Titan Prometheus to the rock of the Caucasus, where an eagle — that is, the reducing activity of the head — bites the liver during the day and regenerates it at night — which is why it constantly sheds bitter 'tears': the bile excretions.

Healthy liver activity leads organically as well as psychologically to a flexible personality with the forces of initiative. If this activity is disturbed then a person 'desiccates'. He may have a great many ideas in his head but they never enter his will. He has no drive and is inflexible, as we can experience at the beginning of a depression, that is also called *melancholy*. One's voice can become brittle and dry, one's being lacks succulence, or in other words humour ('humour' = lat. 'juice'). The muscles, sinews and joints can 'dry out', as can the aqueous humour in the eye. The liver itself suffers the most when it goes hard and stagnates, as we know in the case of liver cirrhosis. In this case chronic colds of the stomach area play a part ('abdominal flu'), when for instance the liver is robbed of its warmth energy through the drinking of ice-cold drinks, so that it 'contracts' and, via inflammation (hepatitis) can assume a degenerative form. But the opposite can also occur when the 'wood' burns too strongly and a red face, overheatedness, irascibility or manic volatility of thought result.

Once you have recognized the Jupiter function of the liver you also understand why the corresponding metal tin is recommended for conditions of illness that have to do with the drying out of the body like arthritis, stiff joints, eczema or depression, where as it were 'buoyancy' is lacking and one can no longer speak one's mind ('speak straight from the liver', as they say in German). This applies also in the case of loss of vitality, the shrinking of sinews and muscles, grey and green glaucoma, indeed, even dryness of the eyes and the mucous membrane; but also in the case of the opposite, oedemas and water retention that the organism cannot penetrate properly: cysts, discharges, and inflammation of sinuses and the jaw.

In brief, we can say that liver activity in the organism regulates the healthy flow of fluids, of 'moisture', so that life forces can gain a hold. The inner connection between the eye and the liver is very interesting. A flooding of the eye by sense impressions harms the life forces, and a disturbance of liver function can show itself in the tendency to tired eyes. If you store up anger and annoyance you can restore impaired liver function by shedding tears. All the functions of the liver show us that they play a dominant role in metabolism. Just as substances that are imbibed are destroyed in the intestines, they are built up again individually in the liver. Especially in carbohydrate-sugar metabolism, the liver is unique in storing the sugar it needs only as a small portion for itself, and, with the pancreas, withdrawing it from the organism when there is a surfeit, or supplying it with more where there is a deficit. In doing this it shows an interesting rhythm dependent on the sun: between two and three o'clock in the morning the storing phase (so-called assimilation) of 'animal' sugar (glycogen) reaches its climax. After this, sugar is introduced into the blood again by the liver, through into the hours of the afternoon, for the day's activity and our mental activities, and this is mainly used, with the help of phosphorus in the muscles, in our expressions of will and in the brain. This also explains why sugar and healthy liver activity are a guarantee of will activity — it is a disastrous idea to deceive the liver with artificial sweeteners! — and why the liver is connected with the heaviness of will in 'melancholia', which sometimes we try to combat in the short term by eating sweets. According to the Chinese organ clock, the 'time of the liver' is between one and three in the morning. So whoever regularly wakes up at this time or stays up especially between 11 pm and 1 am (gall bladder time), or between 1 and 3 am, will, over time, develop metabolic problems to do with the gall bladder and the liver that can appear as digestive disturbances, a lack of regeneration, morning tiredness despite sufficient sleep, a basic depressive mood, especially early in the morning, great thirst or sudden loss of vitality — with symptoms such as hair falling out. While these may not be directly associated with raised liver values, they represent severe functional problems of this organ.

The dominant role of the liver in the body's balance of fluids, normally and much too narrowly blamed on the kidney and its relationship to liquids imbibed, will concern us later on, when we describe the meteorological aspect of the liver. What is certain is that it is responsible for our thirst, as the key water-regulating organ.

In traditional Chinese medicine the liver is associated with the taste 'sour'. Sour foods enliven us, when we are tired, helping us to feel more present, so we can understand that 'sour makes jolly'. Water retention as in the case of cellulitis responds well to external applications of lemon, and these are also helpful added to a bath if we are at risk of faintness from perspiring too much in the summer.

If we return for a moment to the rhythm of the planet Jupiter, we see that it takes twelve years to go round the sun. Applied to human biography this suggests that we should give special attention to the feeling expressions of a young person around the age of twelve. We need think here only of how Jesus presented himself in the temple when he impressed the old scribes with his universal wisdom.

The liver, but also the gall bladder, have an important function in human soul life: in the activity of the will. If a will action becomes too one-sided or even destructive then we speak of aggression or biliousness; if will comes to inadequate expression we speak of weakness of the will or, in an extreme case of depression or melancholia! The will forces in human beings also show themselves within, for instance as thinking activity, while those that initiate outward action are dependent on a well functioning liver activity. Among other things, the liver makes sugar available for mental activity and also for the work of the muscles. This direct material connection, however, is only the external side of a very important energetic relationship: the transference of thoughts into actions. In the case of depression, thoughts get dammed up in the head and paralyse action. This is why bodily activity is always a remedy for depression! In the case of aggressivity, the opposite is true. Here actions erupt without being governed by reflection. In both these one-sided situations one can support the liver and gall bladder with medicinal substances. This relationship to will activity is important since it is symptomatic of people with liver damage caused by drugs, medication, a lot of narcotics, hepatitis or other troubles, that they develop many ideas and plan to do lots of things, but cannot get down to doing them. In this case the liver values do not need to be raised. The damage to the organs is seen chiefly in energy and psyche.

> For the crux of the matter is that the liver is not only the organ in the human being that today's physiology describes, it is in the most eminent sense that organ that gives a human being the courage to turn the idea of action into one actually performed.[29]

So when such blockages occur in the life of will it is because of a subtle liver defect: 'The liver always conducts planned ideas into actions performed by the limbs. In fact every organ exists to conduct or transmit something.'[30] A very important comment that leads us to diagnosis of the effects of organs on the psyche, which Rudolf Steiner referred to as 'intuitive medicine'.

In world literature there is even a famous example of a pro-

tagonist whose constitution — his excessive wealth of ideas — lead
to inaction, so that he chooses his bed in preference to work and
action. This is Ivan Goncharov's masterpiece *Oblomov*. Even his
round, soft and phlegmatic constitution makes one think of a
typical liver case. If someone knows what is needed yet does not
do it, because he is hindered by all the ideas in his mind, he
suffers from a condition that is sometimes even referred to as
'Oblomovism'.

The gall bladder as the Mars organ

We often associate anger and rage with the gall bladder, and this
comes to expression in phrases like 'spitting bile' or 'overflowing
with bile'. The typical tone of voice of someone in whom liver
and gall bladder over-predominate is loud and often raucous.
Through dissolving of old blood, bile is prepared in the spleen,
worked on further in the liver and then led into the gall bladder
where, still yellow at this stage, it thickens to a brownish-
greenish liquid that, in the small intestine, is needed chiefly for
digesting fat. There, the fat taken in with food is emulsified by
the bile acids which the liver has synthesized from cholesterol, so
that it can be worked on further by the pancreatic fluids. The bile
duct and the pancreatic duct usually both converge on the small
intestine.

Given the relationship of bile to iron in the blood (the
haemoglobin) — seen in the fact that the colour of bile arises from
dissolution of the red, iron-rich blood corpuscles — and its
aggressive but essential destructive activity on food, we assign it
to Mars, that is, to iron.

Psychosomatic research has shown that anger and irritation in
particular have a negative effect on the liver and the gall bladder,
and actually change their chemical composition. Vexation as
'blocked will' is psychologically experienced as a severe hin-
drance and holds back the necessary flow of bile. From this
anything from cholestasis to gall stones can arise.

But iron is as necessary for normal psychological and bodily immune processes as is poison for a bee, or the thorn for the cactus. If a person has too little somatic and psychological iron energy his mood becomes dull, he will not be able to protect himself very well against infections, will become pale, tire easily, and swallow without hesitation every insult because he is incapable of asserting himself against the outer world. So it is not a bad thing for a person to be a little bit 'liverish' or, we might say, 'gall-bladdery'. This is also normally apparent in the blood, where a small amount of bile acids always circulate.

I should like to mention a patient here who had to suffer the physical consequence of external restraints imposed on his will. This was a person with many ideas and much initiative, who was always being hindered by the manager of his firm from realizing these intentions. For a long time he swallowed his anger, but then he suddenly developed a serious non-infectious jaundice, for which no organic cause was found either in the liver or the gall bladder. His bile was literally overflowing, and the externally inhibited will, closely bound up with the gall activity, turned against him.

The gall meridian in Chinese medicine runs among other places along the side of the skull and over the temples. Pain or rashes in this area indicate involvement of the gall bladder. In nature we find two very effective medicinal plants for the gall bladder that not only regulate digestive activity but intervene far beyond the metabolic area in the brain and nervous system, and the skin. They are greater celandine and chicory.

We can wonder in what other ways bile — which is distributed throughout the organism — serves us, apart from in the digesting of fat.

> If you had no bile you would be terribly phlegmatic; your hands and arms would hang down, and your head, too [...]. A person has to have bile; bile has to come from the liver. And if the liver is relatively small then a person will be phlegmatic; if the liver is relatively large then a person has a lot of fire in him, for the gall bladder engenders fire [...]. Especially in the case of hot-tempered people, a great deal of bile flows out of the liver, much of which flows into the digestive juices and into the blood.[31]

The heart as the Sun organ

[...] Thus the heart is the fountainhead of life and the sun of the microcosmic world, just as the sun similarly deserves to be called 'the heart of the world'. Through its energy and rhythmic beat, the blood is brought into motion, brought to perfection, nourished and kept from corruption and disintegration. By nourishing us, keeping us warm and giving us life, it does its part in serving the whole body, this god of the hearth, the foundation of life, initiator of all being. The heart of living creatures is the foundation of their life, the prince of them all, the sun of the microcosmic world: on it depends all life, and from it all vitality and energy stream out.[32]

In considering the heart, the most central organ of our life and feeling, we can feel ourselves to be in the best of company with Leonardo da Vinci. 'How shall one describe the heart without filling a whole book?'

In this age of heart attacks, high blood pressure, heart transplants, irregular ways of living and 'heartlessness', it is therefore important that we endeavour to understand more about our organs and realize that the mechanical model of a 'pump' as applied to our central life organ — whose rhythmic beat easily brings it to our awareness — must be critically questioned. Modern investigators are in fact increasingly doing so.

In literature, fairytales, myths, legends and sayings, there is no organ that plays a greater role — also symbolically — than the heart. Proverbially it can be 'broken' or you can have it in your 'mouth' or 'wear it on your sleeve'; you can find something 'heartening' or you can be 'hard-hearted'. The French word for the heart, '*coeur*' gives us the word 'courage'. And in both French and English, the heart is referred to in connection with memory: 'learning by heart' or '*apprendre par coeur*'.

The resilience of the heart's activity exceeds all conceptions. On average it beats about 100,000 times a day, and in a lifetime of 70 years almost three billion times without tiring, despite coffee, cigarettes and alcohol. Daily it has to accomplish a task corresponding to pulling a ten-litre barrel up a tower 200 metres high.

In Egyptian wisdom the heart is seen as the seat of memory,

conscience and the sense for destiny. The reverse of the effect on our heart of shock, deep feelings or trauma is that organic changes to the heart can also influence our soul life in a lasting way. Investigations have shown that after a coronary bypass operation, memory, the capacity for clear thinking and feeling life become less capable, sometimes attended by severe depression and fluctuating moods for more than a year, not to mention changes after a heart transplant, after which serious changes of personality occur. There can also be severe consequences from the administered medicines, ranging from high blood pressure, kidney failure, severe infections and osteoporosis through to a 100 per cent increase in risk of cancer due to the required immune suppression. Up to 15 per cent of those who have heart transplants also suffer cerebral convulsions.

One of the questions most frequently asked after lectures on the inner organs is about the psychological effects of organ transplants. Is it just that one 'lump of flesh' is replaced by another, or is an intrinsic quality of the organ as the seat of soul capacities also transplanted? The changes in character after organ transplants lead us to assume that other effects can be transferred, too, that certainly vary from one organ to another. It is known, for instance, that patients who have had a heart transplant suddenly acquire habits and characteristics, which they did not have before. Research showed that these were qualities that the dead donor had had.[33]

We could add an old legend from South Germany here, which relates to the question of organ transplants in a gruesome though humorous way, yet also provides food for thought. It concerns three soldiers who had left the army, and a doctor. The three soldiers express doubts about the doctor's art. To prove his ability he suggests that when they are asleep he will remove an arm from one, take out the heart from another and the eyes from the third, and put them back in again before they wake, without them even noticing. So this is what he does. When the three are asleep the doctor sets to work and removes the organs. Then he hands them to the innkeeper, telling him to keep them until after midnight. But the cat finds them and steals them away. What is to be done? Where are replacements to be found? The innkeeper remembers that a human heart and a pig's heart have a certain similarity, and so a pig is quickly slaughtered and the heart taken out of it. But where is he to get an eye and an arm? 'Human eyes and cat's eyes are similar,' says the innkeeper, and plucks out the cat's eyes. Then he runs to a gallows and cuts the arm off a recently hanged thief, and keeps all these things in his room. When midnight has passed the doctor appears, asks for the three body parts and puts them back into their old places. The soldiers, who have noticed none of what has happened, praise the great skill of the doctor next morning. All four of them decide to meet again at the same inn in a year's time.

A year later the three soldiers and the doctor meet in the inn again. 'Now how are you, how are things?' the doctor asks the soldier whose arm he had cut off and replaced. 'Yes, it went quite well,' he answers, 'but strange to relate, if I see something that belongs to someone else then the arm that you put back on me tries to snatch it.' This sounds odd to the doctor, and he askes the second, with the pig's heart: 'And how have you been? What have you been doing all year?' 'It's been fine on the whole,' the man replies, 'but if I see any muck I always want to jump in and roll about in it.' 'Strange, very strange,' says the doctor, and asks the third: 'How are your eyes?' 'I am all right,' he replies, 'but for some unknown reason, whenever a mouse runs by I always think I should jump on it like a hungry cat.'

The doctor suspects that something may not be right, and the innkeeper eventually admits what has happened to the three body parts. After the innkeeper pays them a pretty sum of money, the soldiers are satisfied, and as they are otherwise unaware of any problems, they go peacefully on their way again.[34]

But the fairytale does not, unfortunately, relate how they cope in future with their psychological changes.

Now let us turn to an outline of the anatomical heart and its particularities in the human organism, especially the circulation system.

The heart is the first organ to develop in about the third week of pregnancy, outside the embryo. It gradually migrates downward and inward, that is, from the periphery to the centre, initially coming to lie just under the brain's position. This shows us how closely the brain and heart belong together. Anatomists regard this gesture as indicative of a person moving from the 'beyond' into his or her bodily 'presence', the centre of their I being.[35]

In the heart the blood streams from above, below, left and right, intersects, and thus flows through the entire organism. We even find anatomically what is called the 'vein cross' above the heart, while the four-chambered system (the two vestibules and the two chambers) in the heart itself also form a cross. This justifies the notion of a lemniscate in which the heart is, on the one hand, the centre of the smaller upper circulatory system that, via the lungs and the heart, connects us with the outer world; and on the other hand, via the large circulatory system, connects with the inner organs. The heart is the centre of circulation. Within it the blood is compressed to the greatest degree, therefore becoming subject to the force of gravity. On the other hand it enters peripherally into the smallest blood vessels of the skin, the capillaries, under the effect of suction, and thus arrives at levity, buoyancy. When the doctor takes blood from a finger and then puts a glass capillary into the drop of blood one can see how the blood rises up without outside help. The movement of heart circulation is composed of these four dynamic energetic complexes: gravity-buoyancy, and outside-inside. If we now also consider that the heart of the embryo, as a small tube without muscles, already begins to beat, and that islands of blood show movement activity without a connection to the heart, then besides the many additional physiological data such as density, speed of flow, or pressure, we can no longer doubt that the heart, as central organ of rhythm, is not the mechanical origin of the blood's movement. Aristotle already observed, in a chicken embryo, the pulse in the blood islands and described it as a 'leaping point'.

Today we need to shed the mechanistic model of the heart in favour of a living, plant-like model. This sees the heart as a regulating 'retention' or 'banking' organ whose activity originates in blood circulation connected with the peripheral organism, muscles and the need for oxygen of individual organs. Research and experimentation have meanwhile shown this to be indisputable. The Polish heart surgeon, Leon Manteuffel-Szoege writes:

> To summarize, we can ascertain that according to the experiments described in the first part of this book, and the results of previous trials, the author believes there is definite proof that the blood possesses its own movement energy. Supported by our observations it can be said that the blood has specific and spontaneous ability to move within the circulation system, independent of heart activity.[36]

The blood, arteries and heart thus form an inner unity and are connected both with the innermost parts of organs and with the whole peripheral outside world. In the rhythm of the heartbeat and the pulse-wave, is reflected the whole organism. This has been known for millennia in Chinese medicine, with its pulse diagnostics, and is also known in certain investigative cardiac auscultation methods.

Thus the rhythm of the heart expresses the relationship of the upper nerve-sense activity to lower metabolic and limb activities. If the heart beats too fast, as for instance when one has a temperature, or if it slows down, becomes sluggish, as is the case with heart insufficiency, then the dynamic metabolic forces are too dominant. If its beat slows down as for instance in old age, or if it contracts as in the case of angina pectoris, coronary sclerosis or a heart attack, this shows a predominance of head activity, with its focus on breakdown and deceleration. Arrhythmia of the heart, too, is to be seen as a consequence of a lack of coordination between the upper and lower systems. It becomes evident here that, when treating the heart, we also must take account of the other organs and bodily motions.

Let us now turn to the psychological and spiritual aspect of our 'sun organ'.

More than any other organ, the heart, as the central organ of life and as *the* psychosomatic organ, has in the course of the centuries been regarded and interpreted from ever new perspectives. In the same way that the world contains different planes of existence we can identify four aspects in every organ: the anatomical-physical, the vital, the psychological and even a spiritual aspect. This is especially clear in the case of the heart.

As an *anatomical* organ, but not a mechanical 'pump', it is remarkable already in its development and physical structure. The heart chambers with their muscular supports and sinews have even been compared to a 'cathedral'.

We feel our heart as our *life organ* beating continually. A faltering or even a stumbling pulse is experienced as a massive danger, not to mention the feeling of existential threat that arises with angina pectoris or an actual heart attack.

As a *soul organ* and the centre of our breast it is the site of our various feelings—such as constriction, grief, fear, through to psychological armour-plating, but also feelings of love, openness, joy and warmth. We know what it feels like when our heart 'jumps' for joy, or 'pounds' with agitation.

As *spiritual organ* it is the centre of our self and the resonance chamber of our conscience, and intuitively felt truth. 'The heart has its reasons that the mind does not know' (Blaise Pascal). As 'the voice of conscience', as the organ of 'pangs of conscience', as the focus of our sense of destiny and the centre of heartfelt feelings, it has always had a special place in art and poetry. In many a Chinese fairytale and legend the heart has a predominant role relating to memory. One legend recounts that a student who never passed his exams had a new, 'cleverer' heart implanted in him by a judge of the underworld, because the old one 'was not suitable for writing essays, its openings being stopped up'. After the successful implantation of another heart, which the underworld judge had himself chosen from thousands in the underworld, the student made rapid progress in the art of writing, and 'what he had once read he never again forgot'.[37]

From the seventeenth century onwards the 'mechanization of the heart' made headway in Europe, and the dominance of the cool head over the warm heart was established.[38]

In official accounts it was soon being said that, 'formerly we inclined to the heart as the sun, as the king, even, but we find if we look properly that it is nothing but a muscle'. From this time on, the heart was decoupled from the soul as a mechanical pump, ending as a medical problem in heart transplants, pacemaker implants, transplanted baboon and pig hearts and the construction of an artificial heart.

In modern intensive medicine the heart has also abdicated as the organ of life. Nurses and doctors are required to regard a 'brain-dead' person with rosy cheeks and a beating heart as a corpse, and to treat them as such.[39]

But beside the physical heart, and problems caused by lack of movement, stress and obesity, the heart as seat of the soul is gaining increasing importance nowadays. Anxiety, ambition, competitive thinking, suppression of feelings, poverty of emotions, cold-heartedness and also loneliness are known to affect the heart, as do joy, sympathy, openness and warmth. Alongside the concept of heart attack, that of cardioneurosis has also come to be acknowledged.

The hardening of the 'soul heart' (sclerocardia) is seen today by some authors as a significant psychological prerequisite for organic arteriosclerosis of the coronary vessels. Thus the American doctor, Dean Ornish, was able not only to achieve stasis in coronary sclerosis, but even to some degree reverse it by helping patients open their hearts in empathy, perception of their own feelings, the overcoming of egocentricity or recovery of the capacity for sympathy, thus rendering a bypass operation superfluous. This is an approach that has meanwhile been tested and acknowledged worldwide.

The metal corresponding to the heart and the circulation is gold. We think of light, warmth and life as connected with the sun. The sun holds the balance between day and night, warmth and cold, expansion and contraction, we could also say between spirit and matter. A healthy life force, a good blood circulation, the building up of warmth and fortitude are the positive effects of gold. The organism can go off the rails in two directions: with inflammation, in which case the blood 'cooks' in the veins, or with lowered temperature, when circulation falters and a person succumbs to a fear of death. As the heart also has to do with a feeling for destiny, gold can also be used for depression, the 'dark night of the soul', cardioneurosis or high blood pressure, together with the plant of light and warmth, St John's Wort.

Just as the liver acts as a 'general' in the human organism, regulating everything within and needing the aggressive gall bladder process as a 'weapon' to keep away foreign elements in assimilated food, so the heart, with gold, is the life-giving 'sun king' in the organic realm that perceives and encompasses everything, and through which life can 'spread throughout the land'.

If you roll out gold very thinly and then hold it up to the light, you can see that the inner being of gold has to do with the colour of living green.

In traditional Chinese medicine the heart meridian and the meridian of the 'triple burner' end in the little finger and the ring finger. Experience has shown that if one has a cardiovascular weakness one ought to use the little finger and the ring finger either very little or not at all when using the typewriter or computer.

The sense of *taste* assigned to the heart in Chinese medicine is pungent or also bitter, and the *liquid* that belongs to it is sweat. This points to the fact that it is a very good thing in tropical countries to have pungent or spicy food for the sake of the heart and the circulation so that one remains more 'in oneself'.

Later we will examine in more detail the specific relationship of the heart to warmth when describing its meteorological aspect.

We are now going to consider the world of the so-called 'inner' planets. We find the planetary influences of Venus, Mercury and the moon in the kidneys, lungs and the reproductive and regenerative organs. Belonging to *Venus, copper* is the most colourful, warming metal, which among other things is also used in technology as wiring for communication, and in oxidization takes on a vivid green colour. The function—but not the inner structure—of the *lungs* in contraction and expansion belongs to

Mercury; and to the moon belongs *silver*, the metal connected with reflection, renewal and thus with our reproductive organization and cellular regeneration. This metal is found by the ton in dissolved form in the world's life-engendering oceans.

The kidneys as the Venus organ

From Botticelli comes The Birth of Venus, *painted around 1486, an enchanting picture of the goddess of love, who lands on a mussel shell in Cyprus. Cyprus was called this after cuprum (copper) that was found on this island and belonged to Venus as the metal of love. In the middle of the fifteenth century, Botticelli could not know that, according to new biochemical research, mussels and all the other molluscs are copper breathers. We can only be amazed at the kind of prophetic genius he had to have Venus land on a mussel and not on a fish, especially as fishes were assigned to the iron breathers. It was in this way, that, hundreds of years before the discovery of copper*, copper, a tremendously interesting metal in the realm of biochemistry, of life altogether, gave artistic expression to the catalyst of breathing.*[40]

The history of the origins, the function, the anatomy and the psychosomatic significance of the kidneys in the human organism is extremely important. Whoever has occupied himself for a long time with kidney weaknesses or with kidney disease as such, understands directly why, for our whole bodily constitution and the soul life connected with it, care of the kidneys plays such an important role in Chinese medicine — especially in our hectic, stressful modern times, with all their nervousness and irritation. A great deal of what we have to put up with daily literally affects our kidneys and in doing so has consequences that extend to our blood pressure and heart. It is no wonder that our kidneys today are becoming more and more 'porous' and excrete more and more protein, sugar and blood through the urine, if they are maltreated by our modern life style, which scarcely enables us to breathe in a way that is either psychologically or organically healthy.

---

*In the human organism.

In a conversation which the author had some time ago with a Chinese doctor in Taiwan he was told that, in the Chinese view, everything that human beings, and their ancestors too, have ever experienced psychologically, 'lands' at some time in their kidneys and in doing so dictates their bodily and psychological equilibrium. 'Kidney care is actually soul care.' We shall return later on to the problem of today's 'psycho-nephrology'.

Even without specialized medical knowledge, simply through first-hand experience, we know how sensitive the kidneys are situated in our lower back under the twelfth rib, to cold and injury. Kidney colic is very painful! Anyone who has experienced it can understand why the sufferer would like to tear his hair out or jump out of the window.

Now what is so distinctive about the development of the kidneys?

Nature takes three attempts before the kidneys arrive where they are today. The position of the kidneys actually falls during embryonic development from above, from the neck area via the chest realm into the small pelvis and rises again to arrive finally in its present position. Medicine calls these three stages the pre-kidney (pronephros), the archetypal kidney (mesonephros) and the post-kidney (metanephros). A part of the old kidney position fuses together with the reproductive organs. It is quite remarkable that an organ 'falls' from above, that is, down from the realm of the sense organs into the depths of the organism and forms a relationship with the sexual organs, which is why we speak in medicine also about the 'urogenital system'. This reminds us of the descent of the eagle in astrology to later become scorpion, which is said to govern the sexual regions.

The kidneys set up their quarters and surround themselves with a thick 'winter coat', the kidney fat, outside the warm, protective space of the abdomen with its metabolic organs, at the back near the spine — which mostly feels cooler to the touch. Woe betide if this fat liquefies through hunger or illness! Then the kidneys remember their inborn desire to migrate, leaving that spot and becoming either a descending or a wandering kidney.

The region of consciousness in the head with its sense organs is, seen anatomically, the region of the strictest symmetry. In the realm of the lungs this symmetry shifts a little and in the metabolic system it ceases altogether, except for the paired kidneys and the sexual glands, testes and ovaries. So we can rightly assume that the kidneys, like our sense organs and our lungs, have maintained something of this quality of consciousness and have turned into dull, inwardly oriented organs of perception and soul, and like inner 'eyes' look into the vegetative realm of our digestive tract, reacting violently when a person loses psychological equilibrium. Because of the kidneys' eliminating and filtering functions, one can understand why Chinese medicine equates them with busy 'civil servants': they work energetically, examining, making decisions, and showing an exemplary ability to eliminate useless matter so that the superordinate 'state' can function smoothly.

Let us now go a little more closely into the actual functions of the kidneys. Their main task is to purify the blood of salts and everything superfluous to the blood. Similarly to the way they possess an upper and lower part in their overall anatomical structure, so they do too in the fine structure of the small kidney channels looping from above to below and up again. There, by secretion and reabsorbtion, filtering, eliminating, reabsorbing and then eliminating again, a concentration is created that we call urine, which is sucked up again by the cavity organ of the bladder. Permeated by a great amount of oxygen-rich arterial blood, the kidneys remind us, in contrast to the liver—where venous blood predominates—of a semi-conscious organ with the tendency to respire inwardly, like the lungs. Does the kidney also have a hidden inner side that differs fundamentally from the apparent and palpable function of forming urine and eliminating water?

As already mentioned, this is its relationship to oxygen. It is an interesting phenomenon—that we have already seen in the case of the liver in its assimilation of sugar—that more of a certain substance (oxygen in the case of the kidneys) is absorbed than is used up in the organ itself. Only the heart and the brain surpass the kidneys in their absorption of oxygen. In addition to this is the fact that they are in a position, in the case of loss of blood, and high altitude air, if the organism suffers from lack of oxygen, to form a substance that in the bone marrow encourages the production of oxygen, transporting red blood corpuscles (so called erythropoietin). On the other hand, certain illnesses of the kidneys lead to impoverishment of the blood, anaemia. The kidney has a great interest in permeating the organism with air in the proper way, and that means also psychologically. So it becomes understandable that besides its function in eliminating water, nitrogen-containing substances and salts, it regulates the gaseous element in a person, extending its 'breathing' activity as far as the lungs and beyond. You see, it is not the lungs that breathe but a soul-endowed person, and depending on his soul constitution he draws the air deeper into himself when, for example, he is indignant or breathes more easily when he relaxes. In comparison we can imagine the lungs as a purse that only *stores* money. The 'money' is the air that doesn't flow automatically into it, when the 'money purse' lung is opened but must be brought there from elsewhere.

With the help of the cavity organ of the bladder, this dynamic of air forces sucks water out of the body. Although this may seem an unusual idea, this insight has proven very successful in the medical practice of both anthroposophic and Chinese medicine. Regulation of the air organism by way of the kidney organization extends from the intestine through the lungs right into the sinuses and the ears. If we look at the anatomical form of the kidneys we see they have a striking similarity to the ears. The same correspondence is seen in the connection between the urinary tract and the auditory canal, and the bladder and the upper palate.

When infants have an infection, their ears and their kidneys are frequently infected at the same time. It has also been observed that a certain kidney anomaly is often associated with an anatomical distortion of an ear.

As the kidneys, according to the Chinese view, store up the life forces and the inherited energy of one's ancestors, large ears express life force and thus also kidney strength, as we can see in the case of most of the Buddha statues. Although Chinese medicine primarily sees the kidneys as responsible for regulating fluids, as 'water organs', it is also unequivocally asserted that the inner activity of the kidneys draws our breathing deeply into the body and that they are actually the 'source' of breathing. People with shallow breathing often tend to nervous tension, fear or anxiety. Conversely these soul conditions lead to a superficial breathing. Weak kidneys make people more inclined to be fearful, and fear and conditions of shock affect the kidneys, which is why, in dramatic or traumatic circumstances, or in the case of severe injuries, we actually speak of 'kidney shock'. This is where anthroposophic and Chinese medicine meet, and, founded on an extended and deepened knowledge of the kidneys and the bladder function, we can get to the root cause of disturbances in the air organization in conditions such as flatulence, asthma, ear noises (tinnitus) and sudden acute hearing loss, the latter two usually caused by stress, as well as sinus inflammation and ear infections.

Let us look at the kidneys once again from the perspective of Eastern medicine, which calls them 'winter organs' — which, like the earth, store energy in wintertime, so that in the spring, when upbuilding forces are needed again, they can pass it on to the liver, the heart and the circulation. So we especially need our kidney energy when we make too many demands on our forces of regeneration or in convalescence. Thus in the millennia-old textbook of Chinese medicine the *Yellow Emperor*, it says:

> We call the three months of winter the period of closure and storage. Water freezes and the ground splits open. We ought neither to disturb nor impair our own yang* that rests in winter. One should at this time of year go to bed early and get up late the next day. One should restrain one's own wishes and desires as though there were no reason to carry them out, as though they had already been fulfilled. One should keep the cold at bay, looking for warmth instead. Perspiration should cease, and one should fundamentally withdraw from cold energy. All this accords with the

---

* The masculine energy.

lawfulness of winter, and contributes to the maintenance of what one has oneself stored up. But those who do not follow the rules of winter will have their kidneys impaired; the spring will present them with impotence* and they will produce or create little.[41]

'Winter', in our understanding, always also signifies greater density and contraction, as we have seen in the desalination activity of the liver and in the concentration of urine. This is why a salty taste is assigned to the liver. Small or medium-sized amounts of salt stimulate the kidneys, and a hot salty broth is a good remedy for circulation disturbances or low blood pressure. Conversely, too much salt contracts the kidneys, which in the course of time can produce the high blood pressure which can lead further to excessive mineralization, and thus sclerotic conditions. Eye degeneration and general arterial sclerosis can result from this.

As winter organs, the kidneys have to be well wrapped up to keep them warm. Clothing that leaves the kidney area exposed is as far as possible to be avoided, also cold and wet feet. According to the Oriental meridian theory, the kidney meridian goes from the middle of the soles of the feet through the calves and the insides of the thighs, thence to the groin and finally to the collarbone. Feet that easily get tired or that burn, but also chronically cold feet, calf cramps and heavy legs can point to weak kidney energy, which is why hot foot baths have a good healing effect on chronic kidney and bladder infections. As we have already said, the kidney organization and the copper process connected with it support the breathing and therewith the ensouling of the whole organism, so that a weakening of these processes is accompanied by cramps, blockages, limbs and lips that go blue and cold (cyanosis) and general illnesses due to the cold. Venus, with the lower parts of the body, remains as it were 'mermaid-like' in wet, cold conditions. In cases of breathlessness, fear and cramps 'Venus rises up too fast and gasps as it were for air, like a fish out of water'.[42]

Let us now turn to the psychological aspect of the kidneys.

---

* Lack of energy and of productivity.

They are connected with subconscious soul life and, as dully perceptive organs, they influence the emotional soul life and temperament, i.e. the 'wind' in our feeling life. In psychosomatic medicine they are referred to as 'partner organs' that regulate the soul relationship to our environment, in contrast to the heart that influences our relationship to destiny. Thus there is a fundamental difference between something personally affecting and concerning us, and a grave blow of destiny that 'breaks our heart'. However much the individual organs collaborate we find, among other things, that the liver produces a substance (angiotensinogen) which, with the help of the kidneys (renin) is transformed into the substance angiotensin. This in turn affects the muscles of the small blood vessels, thereby raising blood pressure. Some of the medicines used today to lower blood pressure, so-called ACE-inhibitors, intervene in this metabolism and therefore start with the kidneys.

In the case of emotional problems we see how both aspects of the kidneys are affected: elimination of water and breathing. Everyone knows that emotional situations can virtually take our breath away, so that we may almost gasp for breath.

On the other hand it is also well known that emotional occurrences affect urine elimination. We see this in situations like exams and other emotional moments. But conversely urine is retained in traumatic experiences. It is known from psychosomatic medicine that suppressed feelings in children can lead to bed wetting at night, which is why one also calls this trouble 'the tears not shed in the daytime'. If the suppressed emotional problems are released in tears, the increased pressure of urine ceases at once.

In Eastern medicine the kidneys are regarded as the seat of the will, i.e. as the organic foundation for focusing on goals, and realizing them despite outer hindrances.

Modern research on patients with chronic kidney deficiency fully confirms the connection of the kidneys to the soul element. Lack of drive and motivation can be observed, along with resignation, depression and a sense of inner emptiness.[43] The soul is then no longer in a position to take proper hold of the physical body via the kidney organization. A kind of psychological 'becalming' sets in. Conversely, strong and uncontrollable eruptions of anger (storms) and raging emotions can arise.

The kidney organization needs both physical, and psychological and social protection. In the future, heightened perception will be needed both on the part of doctors and laymen to perceive organic weaknesses in the form of emotional disharmonies before they manifest in bodily ailments — when, in other words, they are still at a functional and soul stage — rather than simply shifting everything to a generalized 'psyche' or perhaps to a nebulous 'suppressed unconscious'. The more we come to understand the 'psychology of the organs', the better will we be able to intervene medically on the one hand, and, on the other, the fuller our understanding of others will become, which we need in the social realm for insight into an individual person's organization of body and soul. What, for instance, are the organic and psychological symptoms expressed by a person with kidney deficiency before any manifest illness has become apparent?

Let's take a person who, in his youth, had to undergo trauma and fear, possibly caused by war, and due to frequent relocation never gained any feeling of safety. Or someone whose parents are themselves afraid and literally instil fear into their children's very bones. Then, from an early age, breathing disturbances can arise that do not show any organic foundation in the lungs. The person's mood is likely to be hesitant, cautious and anxious. The feet and sinews are often weakly developed and the feet easily bend or tire quickly when having to do a lot of walking. The characteristic disposition will be gauche, not grounded, flighty, the soul life extremely impressionable to outer circumstances. Whereas a tendency for low blood pressure predominates early on, in old age high blood pressure is more likely to be the case, with strong emotional fluctuations (ranging from transports of joy to deep gloom). The bladder will be extremely sensitive to psychological burdens and there will also be a weakness in the upper air system, with a tendency to inflammation of the sinuses and ears. Stress will easily lead to tinnitus.

If we imagine the kidneys as two curving arms that embrace a tree, thus forming an inner protective space, then we have the corresponding soul gesture. It is a long-held wisdom that people with kidney inflammation should stay in bed for a long time so that the kidneys can recover in constant warmth and with attentive care.

Now why does both traditional, cosmologically-oriented medicine and modern, spiritual-scientific medicine connect the kidneys with Venus and with the metal copper as a medicine?

In mythology Venus/Aphrodite is the goddess of beauty, of spring, and of love. 'Born as sea foam' she is raised up out of the ocean and received in a friendly manner by the winds. This shows her dual relationship to the water and the air, for foam is always a mix of air and water. The basic gesture of this goddess is renewal, of being warmed through, of receiving, upbuilding and soul relationship to one's surroundings. The corresponding metal, red-gold-coloured copper, has properties of heat conduction and malleability, and takes on different colour hues in minerals such malachite or olivenite, just two of the many copper compounds. Copper works medically in the human organism by warming it through, and in doing so it has a connection with the whole nature of the metabolism. The wealth of copper in the liver, our organ of life, also points in this direction.

Some forms of anaemia require copper in order for iron to be incorporated into the red blood corpuscles. It has a special connection to venous circulation and, both applied externally and taken internally, it is a good remedy for congestion, cramps and cold extremities. In the case of calf cramps at night one should therefore not always only think of magnesium.

We have iron and copper in our serum. In the case of chronic inflammations, but also pathological growths as for instance in cancer, copper levels in the blood rise. If on the other hand the metabolic process falters, as in thyroid under-function, then the copper levels decrease.

Copper is interesting also from the psychological point of view. For instance, increases of it have been found in certain 'mental illnesses' like schizophrenia, manic-depressive conditions and also epilepsy. We should also mention in this connection that auditory hallucinations predominate in schizophrenia. This indicates that a connection exists with the kidney organization since, as Chinese medicine says, the renal chi (kidney energy) rises upward and 'opens' the ears.

Let us take one last look at the kidneys with their delicately structured formation of renal corpuscles, which, like receiving hands or blossoms, personify the 'feminine' gesture of this special organ: that of receiving cosmic warmth, love, air and light, helping human beings to ensoul the substances they imbibe.

This is why copper is one of our most important trace elements. A lack of copper leads to osteoporosis and premature greying of the hair. If plants have too little copper they have brown edges and their fruit is either incompletely formed or not formed at all. Copper wire around the roots of tomatoes prevents them from going brown.

The relationship of copper to warmth, including its effect on improving circulation and relaxation of the human organism, was tested years ago at the university of Turin. A cloth of copper placed in the subject's pillow or in the blanket demonstrably reduced head and neck pains caused by stress, similarly to the way copper soles in shoes have proved helpful for a long time for cold feet and feet that easily get tired or swollen.

The lungs as the Mercury organ

We connect our lungs and breathing directly with life and thus with bodily existence on the earth. For everything that is alive must in some way — with few exceptions — take in the 'living substance' of oxygen. We really enter our body with our first breath and leave it with our last. But if we were to breathe only pure oxygen our living substance would soon be exhausted. In fact, together with the air, we take in a large quantity of nitrogen (about 79 per cent), which as it were 'suppresses' the one-sided combustive tendency of pure oxygen, and of which as much is breathed in as breathed out. Along with our heart, lung activity, that is, breathing, expresses an archetypally rhythmic occurrence. Why, in olden times, was breathing sacred as 'the gate to the cosmos' in which the forces of healing dwell? As human beings exhale roughly 18 times a minute, in 24 hours this comes to 25,920 breaths in 24 hours. The sun takes this number of years to pass through the whole zodiac, in what in olden times was called 'the Platonic year'. And the average age of a human being, 72, multiplied by 360 days, comes to precisely the same number—a cosmic 'breath' if you like. In yoga culture, especially, regulation of the breath plays an extremely important role.

Besides our skin, the lungs connect us most closely and unprotectedly to our environment. We are directly exposed and subject to it for better or worse. The environment here enters right into us, and we even have to breathe in what other people have breathed out, causing infections of many kinds in our respiratory tract. In the case of immune weaknesses like Aids, the lungs are the first organs to be attacked.

If we consider the formation of the lungs we see that they are a glandular development of the intestinal wall that migrates upwards during embryonic development, and like an inverse tree implants itself in us with trunk and crown, forming a cavity. It 'sacrifices' its character of glandular secretion to absorb outer air. But when the lungs start to become too glandular again, and secrete too much phlegm, a person becomes ill. As a developed organ, likewise, it cannot deny its affinity with intestinal organization. According to the view of Chinese medicine it has a polar connection with the large intestine, and in anthroposophic medicine we find a reference to the fact that a cough and diarrhoea correspond to one another. We shall come back to this later.

Through the rhythmic movements of their breathing function, the lungs have a relationship to the metabolism and to the brain: during inhalation and exhalation the diaphragm moves up and down, and with every inbreath cerebro-spinal fluid flows round the brain. Thus the rhythmic process of breathing is the real centre of the human being and is positioned, like the life of feeling, between will and thinking.

In breathing, the substance-changing processes that ultimately maintain our life are set in motion in a complex way. Elsewhere we have already seen how our soul life, with the help of our kidney organization, regulates the speed and depth of our breathing. At the same time it is interesting that the lungs are the only inner organ that we can control consciously with our heads. We can hold our breath for a certain time, or speed it up, and can breathe rhythmically or arrhythmically which, of course, has consequences extending to metabolism and circulation. Head forces are what slow down functions in the organism, whereas metabolism makes everything faster. Thus the ratio of breathing to heartbeat/pulse is 1:4, meaning that there are 4 pulse beats to one breath. This ratio varies from one person to another, and can be symptomatic of a discord in the relationship between head and metabolic activity.

Breathing is like a mediator between above and below, the cosmos and the earth, between the damp and the airy nature of the air we breathe, and it is therefore assigned to Mercury and the metal corresponding to it which is quicksilver (mercurius) — that is, the god Mercury who, as a divine messenger, mediates between heaven and earth and is continually in motion on his winged feet. In their organic development the lungs had their own life withdrawn from them so that they could selflessly absorb cosmic life.

All the rhythmic functions in the organism between retention and flow, as in the glands for instance, generally have to do with this mercurial effect. From this we see that all rhythmic processes are half earthly and half cosmic. In our breathing we can experience the cosmic healer of human beings, and therefore we must take special care in childhood that the breathing and thus the lungs develop well. As the activity of breathing is so closely connected with the soul, an intellectual strain is bound to lead to breathing disturbances that make it more shallow. Breathing out also always means dedicating oneself to the world and breathing in is always coming to oneself. If, in early childhood, too strong a 'behavioural corset' is put on the soul, the breath is dammed up in the lungs when breathing out, and this can result in asthma.

Thus care of the breathing becomes care of the soul and vice versa, care of the soul – that is a heartfelt engagement with the world – nurtures healthy breathing.

In Greek antiquity doctors assigned the seven-year phases of life to certain planets. So the first seven years, in which children sleep and dream their way into the world, coming to terms with forces of heredity and mirroring their environment, they are more rounded and watery in nature, under the influence of moon forces. The second seven years, that is, from the beginning of school to puberty, stand under the sign of Mercury. This is when mobility, curiosity, rhythm, the will to learn, and limb activities predominate until, at puberty, with 'earth readiness' and breathing maturation, the Venus age arrives – the time, that is, in which eroticism and sexuality awaken. Between the ages of seven and fourteen, the child does indeed show something of a mercurial nature, as we see in the thermometer: a speedy reaction to outer changes of temperature including in this a certain psychological volatility. The young person gradually begins to come to grips more and more with the earth.

From this it is apparent that a human-scale pedagogy is 'the right medicine' for healthy development of the whole rhythmic organization; an education that does not overburden children with facts and make them inwardly breathless. Let us also mention that, statistically, people are healthiest at this period of development, if there are no soul-injuring interventions from without that, many years later, can manifest as illnesses.

Quicksilver is an important remedy amongst homeopathic medicines, however problematic it is as an environmental poison.

I still remember very well the case of a child with a high temperature who couldn't perspire, and whose temperature wouldn't go down. An injection with homeopathic quicksilver resolved the problem in a very short time. It can be very dangerous if perspiration release does not occur (little 'drops of Mercury' on the skin!) so that the inflammation passes instead into the inner organs such as the lungs or even the brain.

The reproductive organs and silver

When we think about regeneration then the organs of reproduction come to mind as archetype of all renewal in the human organism. An enclave has been created there whereby we human beings are in a position to pass on a new life through heredity. In this specific sphere, millions of sperm cells are formed daily in the male organism to fertilize the woman's egg that has been formed in rhythmic cycles in the course of a month. The average pregnancy of 10 × 28 days is still counted in lunar months, for the moon, the earth's satellite, that waxes and wanes in about 28 days, and is connected with the ebb and flow of tides, is of course related to growth, the woman's cycle and fertility.

The moon is a faithful mirror of sunlight, and it is interesting that in technology silver combinations are used for manufacturing mirrors, and silver salts for photography. That is, the power of silver is essential in a million ways of producing and reproducing images. The greatest accumulation of silver is found dissolved in the world's oceans. Silver seems to us to be an opposite picture to lead. Both of them are commonly found in nature in metal combinations. In the lead process we see isolation, a cutting off, old age and dying life, whereas in the silver process we see thousandfold renewal in cell formation, devotion, youth, and regenerative power — all of this summed up as the enhancement of vitality. Silver is the moon metal that lives in the up-building forces of waters and fluids, and 'Lady Luna' is the mother of boundless fruitfulness. The rising and falling of sap belongs to all growth — as it also does in the sexual experiences of a man and a woman. The moon has always stirred our emotions and been the subject of imaginative and visionary perceptions.

As with all the planets and organs a dual aspect is also present in new moon and full moon. On the one hand the moon has a relationship to the brain, our inner 'dross' that has lapsed from the life process and rigidified, and, like the moon, only mirrors the outside world with no chance of developing a life of its own. If it nevertheless does so, as for instance in the case of migraine, we feel ill. On the other hand, the moon intervenes in our life

forces including the hereditary forces of our reproductive organization. Because of this, the sexual and thought sphere, despite
their polar opposite nature, acquire an inner connection in that
bodily fecundity in the lower region corresponds to the fecundity
arising in the imagination. One encourages the other, as poets in
love have shown us in their works. One instance of this is
Goethe's *West-East Divan*, where we find an inner connection
between procreation and persuasion.

Like all the organs and their corresponding planets, the
reproductive organs have a great many connections to other
organs. The relationship of the sun and moon in the cosmos is
mirrored in the human organization in the woman in the
relationship of the heart to the uterus. Both of them, as rhythmic
organs, are intensively connected with the rhythm of blood and
bleeding, and with feeling life. As the sunlight is mirrored in the
moon and then falls upon the earth, the feeling life of the mother-
to-be is mirrored in the uterus and influences the life of the
developing child. So we have the right to call the uterus 'the heart
of the abdomen'. But the connection between the heart and the
uterus is not only there during pregnancy. Discharge, menstruation, pains, all forms of rhythm disturbances, even muscle
growths (myomas) can be the physical expression of a woman's
unresolved feelings. The author himself has often observed that
serious psychological problems, which persist even after the
ailing uterus has been removed, shift to the next closest 'level',
namely the heart. These connections are also known in orthodox
medicine. There are American and Finnish studies that show that
removal of the uterus before the menopause produces 2.7 to five
times greater coronary heart risk. This is a sign that one cannot
simply remove organs associated with an unresolved soul element without dealing with the actual cause, which is very often
psychological in nature.

We have now given instances of the sevenfold organ processes, of the metals and their planetary connections. One can
imagine what a tremendously rich healing potential lies in
homeopathic metals and their manifold combinations.

Here are these connections once more in a summarized form:

Spleen — bones — lead — Saturn
Liver — tin — Jupiter
Gall bladder — iron — Mars
Heart/circulation — gold — Sun
Kidneys — copper — Venus
Lungs — quicksilver — Mercury
Reproductive organs — silver — Moon

In the same way that we found polarities between zodiac regions, this is apparent also in the planets and the metals corresponding to them. The sun and gold are central here, with their whole relationship to circulation and thus to all the organs. Saturn and lead correspond as dying life and the essential ageing process, and the Moon-silver forces correspond as guardian of regenerative life and youth. Jupiter-tin introduces a forming and life-preserving quality into the organism, and mercury-quicksilver brings everything into movement and dissolution. The male-oriented Mars-iron process has to do with the red blood corpuscles, with strength, activity, consciousness and inbreathing, while the female-oriented Venus-copper process is connected to receiving, nourishing, out-breathing, ensouling and warming. In spiritual-scientific medicine one also calls these the 'seven life processes.'

We now come to a fourfoldness which, as the representative of earthly forces, has to do with the four elements that shape and configure the world. Here we are concerned no longer with the hidden astronomical forces of the planets in the various organs and their metal remedies, but with meteorological forces visibly connected with the earth as we know them in earth, water, air and heat. We can apply these insights in line with the ancient doctrine of dietetics, and the laws governing intentionally elicited healing processes. With the insight that the lungs are connected to the earth, the liver to water, kidneys and bladder to the air, and the heart to outer and inner heat, we are now able, through new, more conscious lifestyles, to make an active contribution to our own health by our own conduct, both as prophylaxis and to address disorders we already have.

We can embark on the path of salutogenesis!

The Pancreas

The pancreas, which in Chinese medicine is regarded also as a unity with the spleen is, like the spleen, a 'boundary organ' that helps us transform the alien stuff that makes its way into us, the food, into our own flesh and blood. To do this the food we ingest must be broken down into fats, protein, carbohydrates, in order to build them up again individually. Otherwise every alien substance would threaten the organism as a 'poison'. In Arabian medicine it was therefore rightly said: We eat ourselves ill then digest ourselves well again. Thus breakdown and synthesis are the prerequisites for a healthy metabolism. The pancreas represents both of these: in the small intestine, enzymes are expelled (or excreted) in an 'outward' direction to break down the fats, protein and carbohydrates, while in an inward direction, into the blood, insulin is made available and intervenes in the metabolism of synthesis. This insulin, originating in the pancreas, creates in the liver a storage sugar (glycogen) from glucose, but there also builds up protein from amino acids, and fatty acids from fats (incretion). This makes the pancreas the most important, superordinate port of call in the metabolism, subject to the ego organization, which we can recognize among other things in the fact that it is the most important sugar regulating organ in the body. For sugar, like fats too, are *the* bearers of our ego and soul organization and are responsible for the creating of warmth, waking consciousness and muscular activity. They can each pass over into the other, i.e., the organism can make fats out of sugars and again make sugar out of fats. An important dietary point for diabetics! Thus in the course of a day about 140 mg of glucose is consumed in the brain. Also more glucose is used in the muscles through bodily work and concentration: heat is created!

Ego activity is both bodily and psychologically connected with setting boundaries. Thus the pancreas in metabolism is the most important representative of inner and outer activity, like the heart in the bloodstream, that must also of course have a physical limit, the cardiac septum, to ensure strict division between outer, oxygen-filled blood from the lungs and inner, carbon dioxide-rich blood arising from combustion processes in the organs. It is an interesting phenomenon that, as with the spleen, excessive psychological self-enclosure can arise by way of the pancreas, so that in certain cases of autism an improvement arises when the pancreas is stimulated by a specific enzyme (secretin).

We see that it is not only purely physical effects that make an organ ill or healthy. The life function of the pancreas can be weakened already in early childhood by psychological 'boundary violations', leading to diabetes, since the border between inner and outer is easily destabilized by repeated emotional upset, shock, intellectual over-exertion, inherited weaknesses, etc. We are then justified to speak in anthroposophic medicine of an ego-weakness in the metabolism. Pancreatic disorders are therefore serious ones: not only diabetes with its destructive effect on the system of nerves and blood vessels (sclerosis) but also cancer of the pancreas, and even its inflammation (pancreatitus).

'Anger sits in the stomach.' This traditional saying should be taken seriously. Will impulses that are not brought to full expression, that were perhaps suppressed in early childhood, or a lack of psychological 'nourishment', can lead later on to weaknesses or even chronic inflammations in this organ. The suppression of the will forces in a strong personality structure, leading through the wrong kind of education or a lack of soul nourishment to a lack of self-confidence, can induce an auto-aggressive, self-destructive tendency in the pancreas. In anthroposophic medicine, depending on the clinical picture, various homeopathic organ preparations made from animal pancreas are available, which, with the addition of metals like silver, barium, and especially meteoric iron act curatively on metabolic processes. We can elicit good effects by using meteoric iron in cases of severe trauma, in states of depressive exhaustion, and also as adjuvant therapy for manifest diabetes.

We actually only understand the activity of the metabolism properly if we regard it as a tremendous process of destruction, that begins in the mouth and culminates in the stomach and small intestine with the participation of bile and the pancreatic enzymes. The psychological outcome of this necessary, inward-oriented breakdown is that one is generally at peace after a meal, through a kind of confining of destructive will forces in our lower realm. But when hungry, one can observe how these forces can actually be unleashed outwardly: in the form of aggression or a bad mood. So having enough to eat makes us peaceful, but hunger makes us aggressive—obviously each person according to type! In conditions where the ego is loosened from metabolism, a kind of 'abdominal clairvoyance' can even arise, as we can see when people take drugs or fast, or also in constitutional weaknesses or an hysterical disposition. In this regard the vegetative nervous system plays a part, especially the stomach brain or also the solar plexus. Applying copper ointment to the stomach area can have a calming, ordering effect and also protect against over-dominant external influences. Thus the egoistic I is situated in the stomach area, and should not be loosened too much. In this connection you can study hungry vegetarians, who are usually very peaceful as long as their needs are met, but, if they can't get the food they need and grow hungry, can become thoroughly egoistic and unsocial. This is not to say anything against a vegetarian or vegan diet, as long as it does not become a substitute religion ... People generally react to their sometimes rather nervy insistence with apt humour: Question: How do you know someone's a vegetarian? Answer: Don't worry, he'll tell you!

The Hormone System

Hormones or so-called 'messenger substances' are formed in the whole organism, for instance as tissue hormones, and not only in the large hormone glands. In the heart, for instance, when we feel pleasure or are happy, the so-called bonding or love hormone is formed, oxytoxin, and affects our emotional brain. Hormones have an immense influence on our whole bodily, psychological and spiritual existence. We can rightly call them material 'messengers' of the world spirit. Saying that someone's behaviour is all 'down to their hormones' is therefore evidence of shallow materialism and should be critically questioned. Although there is no scope here to describe all the complex activities and substances in each gland (that would take a book in itself!) I do want to point to a few important facts that are taught by anthroposophic medicine.

Classically, we have seven major glands which are almost identical to the position of the so-called seven energy centres, the chakras, which all lie on the body's central axis. The hormone glands, from above downwards, are: the pineal gland (epiphysis), the pituitary gland (hypophysis), the thyroid gland (thyreoidea), the parathyroid glands (parathyreoidea), the thymus gland, the adrenal glands (suprarenals), and the female or male sex glands (ovaries and testicles). We noted earlier that the number seven is always connected with the relationship between the psyche (astral body) and the life or etheric forces.

In the case of hormones we find this in marked fashion. Hormone failures are catastrophic for the body and the soul and, even where replaced artificially, always affect both body *and* soul! If we consider them from above downwards, that is from the structuring head forces right down to the metabolic region, we find a striking phenomenon: the pineal gland is small, shrivelled and even has little calcium stones, known as brain sand. In the pituitary gland we find a duality between a nerve-oriented part (neuro-hypophysis) and one oriented to the metabolism (adenohypophysis). In the thyroid gland a threefoldness arises through the connecting bridge (isthmus) between the two lobes, coming to a conclusion in the fourfoldness of the lentil-sized parathyroid glands. Then, in the thymus, adrenal and sex glands, we come to a division into left and right. The thymus gland, chiefly responsible for the immune system and situated above the heart, slowly regresses from puberty onwards and becomes fatty. It has done its job in early youth. It is interesting that 'thymos' means more or less 'soul feelings' in Greek, that is, a heart connection to the world.

The hormones' work is hidden and mostly only two glands come to our physical or symptomatic attention: the thyroid gland and the sex glands. The first of these is in the region of the throat, where today, in our increasingly hectic and nervous times, we must 'swallow' more and more emotional things, some of which may get 'stuck in our throat'. Grave's disease, auto-aggressive inflammations (hashimoto) or other abnormalities such as growths are not uncommon today.

Now what are hormones, in a spiritual-scientific sense?

Hormones are substances, and thus bearers of energy, which maintain youth and vitality in the organism, which otherwise continually tends to disintegrate and is at risk of depositing toxic metabolic products. This is particularly evident in the case of the sex hormones, that are connected with the vital forces of the moon (silver). In a differentiated manner they conduct the cosmic astral into the ether or life forces, which they reconfigure right down to the physical level. Not until the beginning of the twentieth century did people become aware of their outstanding significance. They are bridges between spirit and matter! It therefore makes sense to study the relationship of the seven hormone glands to the planetary forces, the metal processes belonging to them, and the sevenfold etheric functions, as we have already endeavoured to do in the physical realm in relation to the twelve signs of the zodiac. A lifelong but very interesting task!

Before we attempt, by way of illustration, to shed light on the thyroid gland in terms of pathology and therapy, I would like to explain some aspects of the seven life stages in their connection with the organs and glands.

We ought actually also to consider here the 'guardian of the seven', the eighth, which scarcely figures any longer in European folk wisdom. But in relation to the glands, the hypothalamus, a small but superordinate centre in the brain, produces the most important regulatory hormones.[*]

Each of these glands has its own cosmic task and their interplay is very important. The pineal gland is a rudimentary sense organ with brain sand, corresponding to the warmth and mineral forces of Saturn. In us human beings it is active in light metabolism (melatonin). In reptiles it is still an organ for perceiving warmth. In us too it was once an organ for the external perception of warmth, and thus spirit! The pineal gland (lead) ensures a healthy and not too rapid pace of development: if it ceases to work, we see premature development occurring (acceleration, i.e, Pubertas praecox). It is therefore an important brake on the activity of the sex glands and their acceleration of sexual development. Here we see a polarity of Saturn and Moon, i.e., lead and silver. Thus each of the hormone glands helps us understand its opposite picture. Rudolf Steiner connected to the seven levels of life, activities and functions which we could apply to the separate hormone glands:

> Life of the senses — dying life — pineal.
> Life of the nerves — preserving life — pituitary.
> Breathing — forming life — thyroid gland.
> Life of the circulation — expanding life — parathyroid gland.
> Life of the metabolism — material organs — thymus.
> Life of movement — energetic life — adrenal gland.
> Life of reproduction — regenerating life — sex glands.

---

[*] In the mouth we also have an eighth tooth, the wisdom tooth, as the guardian of the seven permanent teeth on each side of the jaw; and even eight hand knuckles. The 'eighth sphere' is described in spiritual science as a very important realm, corresponding in Buddhism to the eightfold path. We can see this in terms of the sphere of the fixed stars standing guardian over the seven planetary stages.

Now let us try to trace these aspects further — not in a superficial or trivial sense — to understand a little more about the inner activities of the gland organs:

- The pineal, with its relationship to the sense qualities of light and warmth and its tendency to build bones. The senses are in general 'dying', that is, they have become physical.
- The pituitary as the gland that rules everything, like an eagle, like Jupiter (Zeus), regulates everything from above, and is a cosmic preserver of all the earthly functions, as is our brain too.
- The thyroid gland with its relationship to the soul element, that is the air, through to its influence on metabolism.
- The adrenal gland as a dynamic factor, which, when it fails, leads to severe adynamy (Addison's disease).

Rudolf Steiner once called the thyroid gland the 'brain of metabolism', and described it as an organ that regulates a person's soul relationship to the outer world. Its substance is iodine, which is capable of destroying and 'combusting' lower life (bacteria), thus facilitating the transformation of vitality into higher soul capacities. If the thyroid gland fails completely then the human being drowns in a sense in life, in water, growing bloated, suffering from idiocy (cretinism) and shutting off from his surroundings. This could almost be described as a plant-like existence. In the case of thyroid over-function, a person burns up matter faster, becomes nervous and thin and his eyes bulge as though they can't get enough of the world (Grave's disease). Thus iodine takes care of our soul development and the transformation of life forces (water) into soul capacities (breathing). For instance, by feeding tadpoles with thyroid gland substance, you can hasten their development into frogs, changing them more rapidly from water animals to breathing land animals! In anthroposophic medicine, there are remedies made from a combination of homeopathic animal gland extracts with corresponding metals, administered as a stimulus for weaknesses of body and the soul. This is very different from hormone replacement therapy!

In cases of thyroid gland growths, an effective remedy is

obtained from part of the meadow saffron, combined with copper and iron. An etheric-cosmic reordering of the organism can be achieved especially via the glands and their corresponding metals.

The Lungs, Liver, Kidneys and Heart as the Four Meteorological Organs

The people of antiquity said that animals are taught through their organs. But I would add to this by saying that human beings teach their organs.[44]

Goethe

Knowing that our inner organs have something to do with the four elements and the four temperaments is not new. People have been connecting the *lungs* as an earth organ with *melancholy*, the *liver* as a water organ with *phlegma* (Greek: 'mucous'), the *kidneys* as an air organ with *sanguinity* (flightiness, airiness) and the *heart* with warmth, that is the hot temperament, *choler*. In ancient systems of medicine people also tried to find an outer analogy in nature for the inner organs of human beings. For instance in Chinese medicine the four organs above are associated with the four seasons: the liver to *spring*, the heart to *summer*, the lungs to *autumn*, and the kidneys to *winter*. For treatment and prophylaxis this gives rise to certain rules of health, particularly obvious in the relationship of the liver, our life organ, to spring:

> We call the three months of spring the period of the beginning of life and of life's development. The energy of heaven and earth is there, ready, so that everything blossoms and flourishes.
>
> After a night's sleep one ought to rise early, wander round the courtyard, loosen ones hair, and only move in a leisurely manner. That is how to lead a healthy life. At this time of day one ought to do justice to the body's urge to live; one ought to be adding to it rather than taking from it, rewarding it rather than punishing it. All of that is in agreement with the energy of spring, and that is the method to apply to safeguard one's own life. Those who do not take into account these laws of spring will be punished by impairment of the liver, and will be afflicted the following summer by a cold illness.[45]

Misconduct in this regard might be seen, nowadays, as something that would weaken the immune system. Perhaps this explains such a thing as summer flu?

The four elements are visible carriers of spiritual effects that are also to be seen in the human being as a bridge for cosmic forces not directly visible: The earthly-mineral element as the bearer of physical forces that in the course of time degenerate; the watery element with its forces of buoyancy as the bearer of all life processes — for without water nothing grows and flourishes on the earth; the airy element that constantly seeks to expand into infinity, as the medium for breathing and language, that is, for the ensouling of the organism; and finally warmth, the 'queen of the elements', which, although it is the most unphysical of the elements yet nevertheless rules and transforms all of them, and can be regarded as the earthly basis for embodiment of the individual spiritual element, our ego. Warmth should be seen as half earthly and half cosmic, which is why we speak of heart's warmth, nest warmth or soul warmth. Every form of devotion, love and sacrifice has something to do with warmth, just as wood has to 'sacrifice' itself so that warmth can be created. The saying has come down to us from Aristotle: 'A human being is born out of warmth, and if he loses his warmth he loses his humanity.' Coldness spells death to all warm-blooded creatures.

The inner relationship of man to nature through the four elements is one of the reasons why for thousands of years medicine has endeavoured to use these four elements to heal physically: with *medicinal clay* or a geological change of location, with *water treatments* as for instance in kneipp hydrotherapy, with *air*, whether using oxygen, carbon dioxide, ozone or even 'a change of air', and with various kinds of *heat treatments*, for instance also by eliciting fever in a patient, which — whether naturally occurring or artificially induced (hyperthermia) — is the arch-enemy of cancer, because it deeply benefits the immune system. In this way we can arrive at a new and above all deeper understanding of what was once called 'geographic' or 'geological' or 'meteorological' medicine. In earlier times people were very aware of the climatic or also specifically geological influences of various soils on body and soul. We find for instance in Paracelsus the knowledge that certain illnesses of the blood occur when the soil has too much iron in it.

The lungs as the earth organ

The study of the geology of a region is actually one and the same thing as the study of the lungs of the region concerned.[46]

'Earth' is a solid mineral foundation, which, though it sustains life, cannot itself produce anything, for the potential for life originates from outside, from the cosmos. Growing and flourishing are not possible without the influence of water, sun and stars.

On the mineral earth human beings take on their earthly mineral body. With 'earth' we connect the idea of solidity, coolness, gravity, form that has become visible, but also being born and dying, work and sweat. The lungs, projecting into us as an upside-down tree, have an average blood temperature of only 35.8° Celsius. This very cartilaginous organ, which, for better or worse, unites us with our physical surroundings, is one through which, by taking in oxygen (life) and giving out carbon dioxide (death), we take hold of our bodies with our first breath and end our earthly existence with our last. So the physical lungs are what first enable us to have life at all; and as we saw, breathing is only a secondary activity of this organ, which on earth brings life into us from outside.

If the lungs are healthy in their physical constitution then they hold an organic and individual balance in us between being too fluid and watery on the one hand and too solidly material on the other. If the inner 'earth' within us dissolves too much, a person will become too airy, flighty, and in an extreme case con-sumptive. If the same element becomes too hardened then the mineral forces dominate, incarcerating a person's inner life so that he may either stagnate or become excessively 'armoured'.

If lung activity tends too strongly towards dissolution, (among other things 'remembering' its mucus-building fluid and glandular character) then bronchitis or lung inflammation occur, marked at critical moments by extremely high temperatures and attendant hallucinations. This can be extremely dangerous for very old people, which is why a lung infection in children must be regarded differently than in later years. If the lungs actually disintegrate into their physical substance as with tuberculosis, there is a tendency to leave the earth altogether, to 'vanish' from it,* which is why it has gone down in history as the 'illness of the Romantics'. In the 1950s there was an investigation into the connection between the psyche and tuberculosis which established that an occurrence of the illness was not the expression of the fight between the organism and germs but between the organ and itself, with an important part played in development of the illness by certain psychological predispositions and conflicts.[47]

We can also consider here the inflammatory and 'dissolute' tendency associated with the smoking of hashish, along with reddened eyes and runny nose. This practice easily withdraws a person from the earth and can lead to illusions or even hallucinations, at the same time to an enormous appetite for food as a counterweight, as it were, to too strong a tendency to release oneself from the earth. Physical nourishment — our daily food — is the most earthly of the things we need for maintaining our bodies. To enjoy eating with a healthy appetite is constantly saying 'yes' to the earth. There are many folk-sayings that speak of the connection of the liver (a water organ) with thirst, but few are aware that, as an earth organ, the lungs regulate our hunger. This in turn explains loss of appetite especially in children after lengthy pulmonary illnesses and lung weaknesses. In the case of anorexia nervosa in puberty, too, as a developmental and maturational disorder, the lungs also require medical treatment. Although this developmental disorder tends towards fleeing the earth, its psychological picture is the opposite, that is, compulsive and controlling. As we already saw, rhythmic activity in the breathing is subject to the mercurial principle, while earthly-

---

*Editor's note: The German for consumption is *'Schwindsucht'*, the first syllable of which means 'vanish'.

physical form and development is subject to the lungs and thus also to the iron process and its principle of bodily engagement.

The lungs' pathological tendency to dissolve, therefore leads in the life of the psyche to flightiness and delusional tendencies as we can read about in Thomas Mann's novel, *The Magic Mountain*, set in a lung sanatorium.

The lungs as earth organ send their earthly energy into our organs of movement that have mostly to do with gravity: the legs. We can also experience this strikingly in the animal kingdom when a tadpole, maturing to a frog, develops lungs at the same time as legs. That the lungs and the energy of the legs, or movement in general, have something to do with one another, can be studied on the one hand in the exhausting perspiration of lung patients after the slightest exertion—and it is for good reason that they need a long rest-cure and should never over-exert themselves—and on the other hand in case histories of conditions that confirm the inner connection of these two earth organs. For instance the psychoanalyst, Horst-Eberhardt Richter, tells us of a traumatic experience during the Second World War when both his lungs and legs simultaneously failed him:

> When, a few weeks later, I entered our half-ruined rented house in Berlin, I found a Hungarian couple in two of the rooms on the third floor, half of which had been spared after a bomb was dropped on our old house. Didn't I know, they asked me, that my parents had perished back in June 1945? How did it happen? It happened in the village where they had gone to escape the bombs. The Russians ... A day after this shock my temperature went up. Inflammation of the lungs. Hospital. I lay there for several weeks. The fever had already long subsided and my lungs were all right again. But I could not or would not get well. I could not stand on my legs. It was the legs once again. I remember the way the specialists and two sisters looked at me critically, and put me through a walking test. I walked very stiffly and was ashamed of myself. I was aware of course that it was not the fault of my legs.[48]

But where are people sent to be cured when they have pulmonary conditions of one kind or another? Away from low-lying ground and limestone soil and to a geological area connected with the strongest formative force: the silica or granite of the high mountains, where the most intense light forces are also to be found, where the 'spiritual aspect' of the air predominates. Or otherwise to where sand (silica) or salt predominate: the sea. It is striking to find how children who, with their close connection to everything watery and flourishing, can't get rid of their bronchitis on a limestone soil, get better ever so quickly when taken to the mountains or to the sea. The inner connection of the lungs to the forces of the earth also suggests why asthmatics improve especially in caves in the ground.

Conversely, though, the lungs can also help the soul when it grows too earthly and compulsive, by freeing themselves organically through inflammation or a haemorrhage from psychological enchainment, obsession or from stagnation. We find an example here in the letters of Franz Kafka, who suffered from tuberculosis and manoeuvred himself into more and more divisive and compulsive situations, that threatened to deprive him of all freedom and movement and estrange him from earthly tasks. His longing to leave the earth became greater and greater.

> My doubts surround every word of mine [...] Everything is fantasy — my family, the office, friends, the street, all is just imagined, whether near or far, (even) my wife [...] The most immediate truth is that you are only pressing your head against the wall of a windowless and doorless cell [...]

But how could he get away from this situation, and who helped him to do so? His lungs produced a haemorrhage, a 'breakthrough,' and by doing so helped him form a resolve:

> The fact was that my brain could no longer bear its worries and pains. It said: 'I give up; but if there is anyone here who cares whether the whole thing continues then let him take some of my burden from me, and things may carry on for a while.' The lungs volunteered, they certainly didn't have much to lose. These negotiations between the brain and the lungs, that took place without my knowledge, may have been agonizing.

The battle for survival by his own lungs became ever more acute:

> What shall two lungs do [...]? If they breathe fast they more or less suffocate from inner poisons; if they breathe slowly they suffocate from unbreathable air, from outraged things. But if they seek their own tempo, this quest alone finishes them off. This proves that it is impossible to live.[49]

So through its illness, a certain organ, in this case the lungs, can help take on and resolve a specific psychological problem. The same is true of other organs as well.

In this regard I once experienced an issue of compulsive-obsessive behaviour in someone with severe anorexia, followed by a lung inflammation. She temporarily lost her psychological symptoms when the lung infection occurred, and the entrenched pattern of self-control visibly 'liquefied'. In clinical practice one repeatedly finds in cases of psychological or also physical trauma that the soul seeks to free itself from its constriction and confinement through 'dissolving' conditions, that is, bronchitis or pneumonia.

But we must also take a look at the other aspect of pulmonary illnesses, that is, in hardening, incrustation and congestions, as we know in the case of asthma. Besides the symptomatic 'holding' of the outbreath we also often find in such conditions a fastidious urge to be clean and the understandable fear of an 'unclean' environment, as well as a fixation on medicines that widen the bronchial tubes. Increasing solidification of the already viscous mucus becomes a torment. Often we can discover a psychological 'corset' that was first established through excessive fixation on a parent in earliest childhood. Marcel Proust, who suffered badly from asthma and never broke away from a close connection with his mother, would be an interesting object of study here, also in respect of his key theme, the 'search for time lost'.

In an asthmatic, the hardened, self-enclosing head-forces risk invading the lungs too. In cases of tuberculosis of the lungs, often associated with episodes of hallucination at the peak of the illness, these structuring head-forces depart from the lungs. We find a typical example in the notes of the writer George Sand, who spent some time in Mallorca with the musician Chopin when he had tuberculosis.[50]

She reports that Chopin often 'got obsessed' with one passage in a composition, and did not relinquish it until he had lung problems.

Concerning the matter of smoking tobacco, the smoke itself is due to a process of incineration, and has the opposite effect to hashish. In the middle sphere of a human being, regarded as the feeling sphere, it has a sealing, consolidating effect apparent especially in the formation of lung carcinomas. This effect, however, can invade the leg arteries in particular as a physical tendency to sclerosis, (so-called smoker's leg), cool blood temperature to about 0.5°C, and assuage hunger, which we can again interpret as a counter-effect.

The healthy centre of the soul, between the extremes of what is compulsive and what is illusionary, between excessive adherence to the earth and excessive release from it, is something we especially need to cultivate, and gives our soul a healthy framework. Here I mean the logic of thought, as we find especially in mathematics and geometry but also in quite ordinary thought structures. What we can call the 'healthy head nature' in our soul is responsible for supersensible forces at work in our lungs, which can deviate in one or another direction if the lungs not only get physically but also functionally ill. Thus the lungs are the 'earth' in us and, as a geological–meteorological organ, are connected with the solidity of each type of soil, and with what we can do only on the earth, namely move under the force of gravity and work with our bodies. These two criteria, the specific geology of where we live and a healthy extent of bodily movement, are of key importance for prophylactics and for maintaining health in a weakened lung organization. There is a note on this by Rudolf Steiner from the year 1920:

> The lungs are like a mirror image of earth conditions—in granite areas they will get well—in limestone soil they will be tested—in sedimentary soils they will tend to fall ill. Lung illnesses are exacerbated by excessive bodily work.[51]

A classic example of premature physical strain and con-
comitant severe lung impairment is that of Anton Chekov, who
died early from tuberculosis. From the remarks above we can see
the importance of a rest-cure and the recommendation to people
suffering from lung damage not to overstrain themselves, or to
change location when they are chronically ill.

So the value of internal or external treatments with silica,
granite and salt compounds is clear, as well as, in anthroposophic medicine, a remedy containing an ash preparation of
certain medicinal plants. For ash is the earthly residue, the pure
mineral part that alone remains after cremation. Since olden
times plants with a marked relationship to the earth — for
instance the bitter plants, Icelandic moss and wormwood — have
been successfully used in herbal remedies.

> If someone is as green as a tree frog, as thin as a poplar, loses
> weight and humour everyday, and scarcely casts a shadow, try
> taking a spoonful of wormwood every two hours, for it is not only
> an excellent remedy for lack of appetite, but also for purifying the
> lungs.[52]

In their relation to the earth, in some patients with severe lung
symptoms the lungs sometimes even produce musty odours, and
because the lungs are the organ of grief we often find a tearful
tenor of voice manifesting especially at moments of agitation.

We mentioned earlier that there is a functional relationship
between the lungs and the intestines, or, seen symptomatically,
between a cough and diarrhoea. It is known in practice that
sometimes when a cough is beginning to get better, diarrhoea
can occur, or in cases where the cough is severe, an enema can
help improve it. This knowledge leads in many cases to con-
ducting an intestinal purge. We also know from medical practice
the problem that symptoms occur suddenly in the intestines if
we suppress the lung illness, or vice versa.

The lungs separate during embryonic development from the
primal intestines and move upwards. So we can discover a
functional contrast between the lungs and the intestines but also
a similarity. There is also a resemblance in the fact that they both

excrete superfluous, toxic matter, either in the form of air, or solid matter. They are opposites in the fact that what is healthy for the lungs becomes an illness where the intestines are concerned, and the other way round. That is, the intestines produce large amounts of mucus and do not especially like air, whereas the lungs enjoy filling themselves with air and may only produce a small amount of mucus, but tend towards illness if they do so excessively.

In Chinese medicine, besides the relationship of the lungs to the large intestine, and to the hardest thing on earth, 'metal', there are also interesting concordances with the spiritual-scientific view of the lungs as 'the earth' in us. In Asia the lungs are regarded as 'the minister' who, in the harmony of the circle of organic functions, is in charge of order in the physical 'realm', and constitutes the earthly 'surface' for properly assimilating energetic influences from without, such as light and air. According to the Chinese view, the lungs take care of the organic energy needed to construct our bodily house. They rule the 'defensive energy' of the skin as one of the most important organs of immunity in its creation of sweat—we think here immediately perhaps of the exhausting night sweats of lung patients—and by warming the entire skin surface. According to the Asiatic view, not only moistening but cooling of the whole bodily system originates with the lungs. From this perspective is it not surprising that one can, for hours, breathe in cold air at minus 40° Celsius without getting a lung inflammation? But if, by contrast, the abdomen, especially the leg and kidney region or the feet, get cold and wet, then in no time the 'upper regions' will show bodily symptoms of a cold.

The emotions connected with the lungs are grief and sadness, as is, therefore, the melancholic temperament. These draw us most strongly earthward. Corresponding periods are those when the day or season are 'declining': the afternoon or autumn with its earthly smells.

As the lungs are the centre of a human being, lung patients ought to pay attention to their diet, and eat a lot of leafy vegetables. Onions with their sulphuric sharpness also belong here, and the diverse kinds of cabbage, that are primarily leaf. If the lungs tend to be inflamed then we recommend stimulating them by giving them the formative and shaping forces of cooked roots such as carrots, and especially beetroot, which also have a favourable influence on the immune system.

The liver as the water organ

The liver, as our inner 'general', promotes and maintains our life forces. Life has to do with the cosmos, with sunlight and sun warmth and their earthly bearer is water. If the metabolic organism lacks inner warmth then it is a good thing either to put a hot water bottle on the liver area or to expose the abdomen to the sun. If an infant after birth, for instance, has too much bile in its blood (hyperbilirubinemia) due to a disturbance in the liver's development, the child should be exposed to sunlight so that the bilirubin can normalize. Thus sun activity enhances the liver activity, and bile is the earthly substance that 'burns up' deposits both in the metabolism and in the soul.

Our thirst, including the intake of liquid during the day, is determined by the liver, with the help of an important protein substance (albumin) that it forms. This governs the exchange of water between the blood and the tissue, transforming 'normal' water taken in from outside into 'living' water belonging to the body. The liver is actually to be thought of as a sucking blood-sponge encapsulated in a tissue-like covering. Its dominant function in the body's water balance also gives the soul 'buoyancy'. So it is understandable that on the one hand the liver is dependent on clean, vital water and on the other hand should not be overtaxed in its detoxifying metabolic function, as everything that we ingest in the way of food has to be passed through the liver. Water containing limestone is an especial burden on the liver, for limestone has an organic connection with hardening and death processes. If, for instance, thinking is overstrained by abstract thoughts for a long time, then a cooling of the whole organism can follow, right down to the feet. During state exams I saw two of my fellow medical students developing hepatitis because of excessive intellectual strain and nightlong study.

Another one-sidedness that the liver cannot tolerate could be described as 'gourmandizing'. When an unrestrained appetite asks too much of the liver, it tends slowly to degenerate, which occurs for instance in the condition 'fatty liver'. This not only indicates the need for moderation but also conscious chewing that properly relishes food in the mouth, and good seasoning of foods. These stimulate the liver to healthy activity. The liver must not be overtaxed by a surfeit of things such as excessive fat, alcoholic drinks and chemical and substitute sugars such as aspartame.

Major medicinal plants for the liver include, especially, ones that contain lactiferous or sap-rich substances such as dandelion, bryony, celandine and milk thistle, or very sour ones like barberry. Chicory is especially helpful in the case of liver and gall complaints.

The kidneys and bladder as air organs

At an early stage of embryonic development, breathing and nutrition are still a unity. The embryo floats in the nourishing amniotic fluid which serves as both its source of nourishment and 'respiration'. Later on, breathing air, the lungs grasp hold of the formative forces of our cosmic environment, while the liver is responsible for earthly nourishment, the chemistry of material synthesis, and the heart and the circulation take care of the diffusion of air and substance through the entire organism. The kidneys, as a soul organ, synthesize and individualize the substances ingested from without in order to build up the body, and configure the flow of air to become part of the soul, that is, interiority. Thus the kidneys become the organ of 'subjectivity'. The greater the need for oxygen, and the deeper our breathing, the more thoroughly does the soul element engage in our body and combust assimilated substance, transforming it into energy. Interestingly enough, as far as the need for oxygen is concerned, the kidneys and the cerebro-spinal system are primary, and thus turn out to be part of our organs of consciousness.

Breathing, which, as we saw, is organized via the lungs by our kidneys and bladder organization, is dependent on a person's soul constitution. If we become subject to inner or outer strain, i.e. stress, or if we are nervy or otherwise psychologically 'breathless', or if we begrudge ourselves sufficient peace to eat, or bolt our food down thoughtlessly without chewing properly, the inside motions of the organism become chaotic and impair the bladder and the kidney in particular, leading often to an accumulation of air in the upper stomach. The person begins to swallow air and their whole breathing becomes irregular.

So it is of the utmost importance for the life of the bladder and kidneys to maintain the right 'breathing rhythm', especially when eating. This is greatly neglected today with modern eating habits such as fast foods or use of microwaves.

We can enhance the organism's healthy perfusion with breath by taking people to sub-mountainous regions where the air contains more oxygen; and they then unconsciously inhale and exhale more deeply. Irritations of the bladder and the kidneys, manifesting in a burning sensation (irritable bladder), more frequent urination through to volatile fluctuations in blood pressure, improve, when patients learn to breathe more deeply and consciously. A patient who suffered from unstable high blood pressure and difficulties in breathing once told me how helpful it had been to breathe regularly in and out when snorkelling. In this respect one can justifiably say that care of the breathing is care of the kidneys!

The Chinese bladder meridian runs from the inner corners of the eye and nose through the sinuses along the temple into the occipital area at the back of the head, and through the neck muscles, the back and the calves into the little toe. Patients who suffer from weather-induced headaches — especially low pressure or thunderstorms — have problems especially along the bladder meridian, with sensitive places or pain originating in the eyes and passing over the temples through to the region where the musculature reaches the occiput.

So we can also start with the bladder when seeking cures for disorders in the air organism such as sinus inflammations, or lymphatic growths. Such conditions are also associated with an increased need for oxygen and over-sensitivity to light, smells and noises. Similarly to the way in which the liver supplies the eyes with vitalized water, the kidneys supply the eye and the skin with light. This also involves the relationship between the kidneys and the adrenal gland, which co-determines vegetative processes by forming cortisone and adrenalin, two substances belonging to the body that can be used medicinally in cases of shock and severe forms of allergy. Patients with disorders of the kidneys and adrenal glands eventually develop a dull, dirty, grey-brown skin colour.

Medicinally, the air-filled horsetail, which is chiefly stalk and has a high silica content, is used in cases of kidney and bladder infections. Also burdock root as a tea, an ancient folk remedy, works not only on the kidney but externally, used as an oil extract, can also have a beneficial effect on the hair.

The kidneys are nitrogenous organs, so they benefit from plants in our diet that contain nitrogen, like pulses, though these can also make them somewhat dull and cause internal 'wind'.

If the kidneys and the bladder are disturbed in their eliminating function—Chinese doctors speak of too strong a 'contraction'—melons and pumpkins can assist the passing of water and dispel breakdown products from the body. As the kidneys possess a special relationship to common salt, it is very important to limit its intake so as not to overtax them. On the other hand, cooking salt as a homeopathic medicine (natrum muriaticum) is an excellent remedy in the case of a number of kidney symptoms that manifest in an increased need for cool, fresh air, oversensitivity to the rays of the sun, heat, overheated rooms and a fear of uncertainty. In addition, the homeopathic picture for this medicine shows a tendency to moral stringency and an extreme cleanliness compulsion.

The heart as the warmth organ

The heart is the centre of all movements, and it is well known that warmth arises only out of movement. In Chinese medicine the heart has an inner connection with summer.

> The three months of summer are what we call the period of sur-feited growth. The energies of the heavens and the earth unite, so that everything blossoms and brings fruit. After a night's sleep one ought to wake up early. During the day one ought to be calm and composed, and allow development and flourishing life its free rein; one should let one's own energy come into contact with one's surroundings, and should behave as though one loves everything around one. All of this happens in accord with the atmosphere of summer, and is the way to safeguard one's own development. But those who do not abide by the laws of summer are punished by an impairment of the heart. They are affected by cases of malaria in the autumn,* and therefore little energy remains for them to cope with autumn in a healthy manner, so that they can be overtaken by serious illnesses in the winter.[53]

We know that people with heart trouble often suffer greatly from the summer heat and that in the case of the heart being 'overheated' cool poultices in the heart area, under the arms and over the pulses — over which the heart meridian also runs — have a calming effect.

The heart has a connection with the *outer* movements that have to do with our conscious I and less with the unknown nature of soul life, as we saw it in regulation of the *inner, soul* movements in their relationship to the kidney and the bladder.

By means of ensouled, conscious movement one can send warmth into limbs that have become cold, as happens in the case of autogenic training. I well remember a patient who suffered all the time with cold hands and despite physical measures and medication gained no improvement. When I met her again many months later and gave her my hand I was astonished how warm her handshake had become. To my question as to the cause of the

---

* In western medicine we would say severe infectious.

improvement she replied that for the first time in her life she had taken up something that had really made her happy and enthusiastic. Popular wisdom has long known that cold thoughts lead to cold hands.

The heart as the basic organ of our rhythm and circulation is situated between the mutually determining head pole of perception and the metabolic pole of movement. If our harmony is disrupted by sensory overload with a simultaneous inaction of the legs, as with driving a car or using the computer, then the heart and the circulation inevitably suffer. The opposite is true in one-sided engagement in sport. This is confirmed by cases of sudden death caused by heart failure among professional athletes. Modern heart physiology has recognized that the most balanced and best form of movement is walking or hiking and not over-energetic jogging. Also, after a heart attack, it is recommended nowadays — unlike previously — that one should get on one's feet again soon and be active. We can now understand why sitting down for a long time, or much travelling or flying, make heart illnesses worse.

Doing soulful movements, as is done in the eurythmy therapy of anthroposophically extended medicine, is the best therapy and prophylaxis for heart complaints. These are movements 'in which the muscles celebrate a festival' to use Nietzsche's expression.

As early as 1920, long before heart and circulation illnesses increased to become primary in illness statistics, Rudolf Steiner spoke about the fundamental significance of movement for the health of the heart. I will quote it at length here because of its topicality:

> I would particularly like to draw your attention to the fact that you should really endeavour to attribute every kind of heart damage to a disturbance in human activity. You really should investigate how different the heart activity of a person is who, let us say, cultivates his fields as a farmer, with little respite from this activity, from people who for instance have to do a lot of driving, or even only have to travel by train a lot. It would be extremely interesting to investigate this thoroughly — and you will find that

the tendency to heart disease is essentially due to the fact that a person, whilst being moved about by outer means, himself sits still, that is, sits in a train or a car and is being ferried about without his involvement. This passive giving of oneself to outer movement is what disrupts all the processes in the heart.

Now all the factors of these kinds in human life affect the way in which we grow warm or not. And here you see the relationship of heart activity with the impulse of warmth in the world, and its relation to human life. When someone creates enough warmth through their own activity this specific measure of warmth they develop in their life processes is at the same time what is required for the health of the human heart. In the case of heart patients we therefore always have to ensure that they engage in movement that properly and fully enlivens them.[54]

For modern medicine it is self-evident today that active movement alongside nutrition represents the best prophylaxis for circulation disturbances. Inner and outer permeation of the organism with warmth is an important matter in the case of circulatory disorders. Stimulating skin warmth by energetic rubbing or the external application of warming plants like camphor and rosemary (also drunk as a tea) has long been a well-tried remedy for heart and circulation disorders and the tendency to faintness.

The colour corresponding to the heart in Asiatic medicine is red, which makes us think of the colour of one of our best remedies in herbal medicine, the red fruit of the hawthorn.

As we already mentioned, the appropriate season for the heart is the summer, which is why in the case of 'overheating', such as chest tightness or racing heart, we won't go wrong if we calm the heart with cool poultices – in contrast to the liver, which, except in a few exceptions, should usually be treated with warmth.

The emotion corresponding to the heart is joy, and along with this the heartiness we so greatly lack today in this 'cool' age.

The bodily fluid assigned to the heart functions is sweat, which, for Asiatics – who, by the way, hardly perspire at all – is regarded as a valuable bodily fluid that should be 'retained'. When heart energy has become weak then people sweat in their sleep. In the Chinese view, the tastes corresponding to the heart, and good therefore also for circulation, are sharpness and bitterness. Both are very helpful when staying in the tropics or going on tiring journeys by plane or car; they ease the function of the heart via the digestion, since the small intestine is the sister organ of the heart.

At this point I would like to add as a postscript an important research finding of recent years concerning the human heart. In 2007, a working group led by Professor Sengupta in Rochester (Minnesota, USA) published important research findings in the periodical *American College of Cardiology* concerning the flow dynamic of the blood in the left ventricle of the heart. Using ultrasound they were able to demonstrate that the blood forms a vortex, which causes the heart to contract. This conclusively put to rest the old dogma that the heart is the originator and cause of circulation.

In addition it was repeatedly observed that when, in line with the common pump theory, we reinforce left-heart activity with medication, in the case of the heart insufficiency (which, if the heart *were* a pump, would be the sensible thing to do!) this leads to an increase in mortality rates. We know today that an improvement will occur if we reduce the heart's resistance to the blood stream (which of course impels the heart). This has led to the modern treatment using beta blockers, ACE inhibitors, anti-water-retention measures etc., which help reduce resistance to the blood flow. In other words, the treatment approach now employed runs completely counter to the mechanistic model!

If we compare statements by Rudolf Steiner about heart activity with the recent physiological findings, we can marvel at the relevance and accuracy of the former!

I will quote firstly from a lecture to the workers in Dornach:

> The blood wants to receive nourishment (again). The blood as it were draws to itself the food that the stomach and the intestines have imbibed. All this, this hunger for air, hunger for food, brings the blood into movement. It is the blood that moves first and the blood takes the heart with it. So it is not the heart that pumps the

blood through the body but the blood that moves because of its hunger for air, hunger for food. And that is what moves the heart...[55]

What initially sounds so innocuous and naïve here, is confirmed by the wisdom of Chinese medicine, at least in so far as the small intestine is the polar organ to the heart, and is responsible for the assimilation of food. And today modern physiology fully affirms the participation of metabolism in blood flow. It has been found that the blood flows all the more quickly the more oxygen is combusted in the tissues, that is, the greater the metabolic activity!

> Thus the cause of blood flow is always found in the same place where blood volume is engendered — that is, not centrally in the heart, but everywhere in the organism's periphery.[56]

But in relation also to psychological occurrences such as loneliness, death of a dear one, depression over an unhappy love affair, etc, interesting results were recently discovered, such as the syndrome of a 'broken heart'. Heart catheter investigations where heart attack was suspected found no constriction of the coronary vessels but a strangely rounded deformation of the left heart. In ultrasound images this appears like a clay vase or a Japanese squid called a 'takotsubo'. Thus the condition is also called 'takotsubo cardiomyopathy' or 'broken heart syndrome', nine out of ten cases of which were found in female patients. The heart is swamped by stress hormones (adrenalin and noradrenalin) that rise by 35 per cent, as with a heart attack. A few months later the heart can regenerate itself again without any impairment. The chief causes of a broken heart are severe blows of destiny, the death of a loved one, an accident, an attack, severe threats, etc.[57]

The insufficiency of the right heart is a big enigma in orthodox medicine, and at the moment there is no rational treatment for it. The best results are achieved by naturopaths and anthroposophic physicians with a remedy that, thank God, has come to the fore again recently, and that both hinders a heart attack and also works to improve blood circulation. Rudolf Steiner spoke of this—strophanthin or strophantus, an African climber—as a remedy for the damaging effect of modern civilization on the heart.[58]

Poisoning and Detoxifying

If smoke is coming out of the chimney and you want to stop it you will hardly climb up onto the roof and try to catch it there. No! You know that the smoke is coming from the fire that is burning in the hearth; so you will let the fire go out, then the smoke will disappear.

Now natural medicine doesn't run after the smoke, but seeks to put the fire out: in the case of illnesses we do not exclusively treat the organs in which the illness comes to expression but look for the causes of the trouble and endeavour to remove these. The causes are not usually in the affected organs themselves but, for instance, in the lack of activity of the eliminating organs; that is, primarily, in the kidneys and the intestines.[59]

Johann Künzle

It has always been the endeavour of traditional western as well as oriental medicine to provide the human organism, via food, air, water and medicines, with substances that burden human life and activity as little as possible. Ancient dietetic rules of health about assimilation and elimination originate from this. Every intake of substances from outside — including light, air and warmth — was principally regarded as a 'poison', as something foreign to the organism, that the 'inner physician' of our digestion transforms into our own flesh and blood. In other words, these substances are incorporated, and what cannot be used is eliminated by organs that are there for the purpose, to avoid burdening the organism unnecessarily, thus making it ill. This is the foundation of the oriental wisdom that 'we eat ourselves ill and digest ourselves back to health'.

To speed up this process of elimination, the greatest value is attached to sweating, and blood purifying cures, bloodletting, fasting and purging the intestines.

But in the course of human development, our organism has had to come to terms not only with natural substances but increasingly, today especially, with synthetic aromas, sprays, fertilizers, and chemical additives of all kinds deposited as foreign mineral matter in the organs, hardening the living organism and making it impervious to cosmic life forces. Amongst these are the chemical 'attacks' of vaccines with their aluminium and mercury additives, temperature-lowering drugs, antibiotics, vitamin D, fluoride and excessive consumption of animal proteins.

But in our natural metabolic and breakdown processes, small quantities of toxins are also created, especially in the breakdown of protein—substances such as cadaverine and putrescine (from *'putresco'* = to go bad), which the organism then eliminates via the liver, gall bladder, kidneys and the intestines and, if need be, also by way of the skin and mucous membranes. We can also think of the natural formation of alcohol in metabolism and poisonous carbon dioxide that is breathed out through the lungs. In severe illnesses such as tuberculosis and cancer, or severe burns, people can also be poisoned by their own decaying protein.

However, the destructive action of toxins not only affects the living organism but also alters our consciousness too through breakdown processes in the body. The effect of recreational drugs depends on this: life is sacrificed to engender altered consciousness.

It is well known in traditional schools of medicine that, to maintain itself, the organism always works from inside outwards, that is, centrifugally, to eliminate what is injurious to it, sometimes even through suppurative processes. For instance in naturopathy, additional laxatives (!) are given in cases of non-infectious diarrhoea—which is mostly a symptom of food intolerance—to support the organism in its detoxifying activity. This might seem an almost absurd idea in the era of 'blocker' and counteractive treatments. But suppressing the elimination process, in a centripetal manner, will make the illness worse, or move it to another organic level.[60]

Poisons can also serve human beings. I am thinking here not only of enjoyable poisons like coffee, tea or tobacco that cheer our lives, but poisons from the natural kingdom, which, if properly prepared, and given, can be helpful in cases of life-threatening psychological and bodily emergencies. The German word for poison, '*Gift*' also means a 'present', as it does of course in English.

In homeopathy in particular, but also in various other medical systems, mineral, plant and animal poisons are used very successfully in therapy. The poisons that are transformed into remedies stimulate the strongest life forces in people, and by this means the organism can strengthen its activity and recover.

Whether we can physically detect various toxins or substances in the organism or not, they are present as energy potential and must be raised by the human organization to a higher level so that they relinquish their one-sided nature. But if the organism succumbs too greatly to natural activity, and if the one-sided tendency of mineral, plant or animal becomes too strong there, as can be found in cases of kidney or gall stones, the growth of bacterial flora and fauna in the organs, or in the 'animalization' of the soul life, then illness occurs. Something lapses in us from harmony and equilibrium, and becomes subject to destructive natural forces. Thus Paracelsus for instance does not speak of a person 'suffering from cholera' but of a person in whom dominates the natural arsenic-forming activity, which is why he called him an 'arsenic person'. Arsenicum homeopathically prepared as a remedy, that is, 'potentized' as Hahnemann understood it, introduces from *without* the same pattern of illness into the organism against which the organism then attempts to protect itself. In overcoming the arsenic externally introduced, the patient overcomes his own pathological arsenic activity, in this case cholera. The homeopathic principle of '*similia similibus curantur*' is based on this — likes are cured by likes — although this is not its only healing principle. Homeopathy uses natural substances to create empirical, artificial poisoning 'pictures' so as to discover what a 'belladonna illness', 'phosphor illness' or 'bee poison illness' looks like, which one then attempts to cure in a genuine case of illness with the corresponding potentized remedies, calling upon the organism to cure itself. To illustrate this we could think of the psychological healing process of someone entangled in problems, with consequent depression, who encounters someone in whom he sees the same problems as he has. In trying to help this person, he solves his own complex and overcomes his depression. We can also see this phenomenon in self-help groups or when two people of the same temperament 'collide with one another'.

We know that there are substances in nature, in the mineral realm, the plant realm and the animal realm, that are poisonous to us. Examples spring to mind such as the heavy metals like phosphate, nitrate, lead or cadmium, which tend to accumulate in fungi or in animal livers. Furthermore we know the classic poisonous plants or substances that arise in food that has gone bad such as mould fungi and bacteria or, in the animal kingdom, usually harmless insect poisons which can, though, become extremely dangerous in allergic reactions, not to mention snake or toad poisons, or illnesses caught from animals such as rabies.

In cases of acute life-threatening poisoning we must of course resort to specific medical intervention, but it is good to remember that there are also self-help measures in either emergency or chronic circumstances to support the medical detoxification interventions.

Mineral poisoning

Minerals or metals that we call trace elements are generally vital for the organism, and one should preferably absorb them from vegetables or as a food supplement made with organic substances. There are sufficient amounts present in organic and especially in bio-dynamic foods.

Today we must ask ourselves whether the 'American' custom of swallowing enormous amounts of artificial vitamins or trace elements really helps the organs or whether in fact we are forcing them into mineralization and thus into excessive density. In this way *physical* processes gradually acquire the upper hand and thus hinder the life forces and also the soul from properly engaging in and organizing the body. We could think of this contracting and densifying process as an artificial 'glass coffin' that consolidates the organs and inures them from responding to the finer effects of remedies, since a mineral, dead 'phantom' takes up residence in the organs. This will especially be the case when taking psycho-pharmaceuticals, almost all of which originate in the petroleum and chemicals industry. Such medication may well be necessary sometimes of course, but it is still worth making a thorough study of this process of compaction affecting both the psyche and the body.

Back in the nineteenth century people still had a healthy sense in the USA for the fact that one shouldn't simply stuff mineral substances into people when they are ill. In one of his novels Herman Melville described a scene on a Mississippi steamer in which a herb doctor attempted to enlighten a patient about the 'mineral doctors' and the effects of what they were doing:

> You tell me, that by advice of an eminent physiologist in Louisville, you took tincture of iron. For what? To restore your lost energy. And how? Why, in healthy subjects iron is naturally found in the blood, and iron in the bar is strong; ergo, iron is the source of animal invigoration. But you being deficient in vigor, it follows that the cause is deficiency of iron. Iron, then, must be put into you; and so your tincture. Now as to the theory here, I am mute. But in modesty assuming its truth, and then, as a plain man viewing that theory in practice, I would respectfully question your eminent physiologist: 'Sir,' I would say, 'though by natural processes, lifeless natures taken as nutriment become vitalized, yet is a lifeless nature[*], under any circumstances, capable of a living transmission, with all its qualities as a lifeless nature unchanged? If, sir, nothing can be incorporated with the living body but by assimilation, and if that implies the conversion of one thing to a different thing (as, in a lamp, oil is assimilated into flame), is it, in this view, likely, that by banqueting on fat, Calvin Edson[†] will fatten? That is, will what is fat on the board prove fat on the bones? If it will, then, sir, what is iron in the vial will prove iron in the vein.' Seems that conclusion too confident?[61]

---

[*] Prepared as medicine.
[†] The name of an anorexic man who, literally having been reduced to only skin and bones, was being passed round by a showman in New York.

These remarks shed light on the secret of nutrition and the transformation of non-human substances into one's own flesh and blood, a process that Paracelsus described as the highest form of alchemy. For who can even imagine how a cow sets about turning grass into milk, a substance, which, by itself, maintains life and can be refined into numerous further products?

Normally we have a detoxifying mechanism in our organization, an 'inner physician' who helps us cope with the poisons we have ingested or which have arisen in the metabolism. For instance, poisons dissolved in fat are deposited in our own fat and are therefore rendered harmless or have their harmful effect mitigated through biochemical processes. Heavy metals can then only be mobilized if they are transformed into an organic compound and assimilated again into life processes, subsequently being eliminated via the liver and the intestines. A substance called glucuronic acid, derived from sugars and containing phosphor, plays an important part in this. In higher doses, Vitamin C (ascorbic acid) can partly take on a similar function.

In principle the organism's own protein has a detoxifying action, especially in the case of minerals and heavy metals. In cases of poisoning a detoxifying effect can be brought about by giving chicken protein, milk or even plant oils. These substances are the bearers of strong vital forces, and can resist the attack of purely mineral actions. Particular oils, such as sunflower oil, are used as a morning mouthwash for about 10 to 15 minutes, to rid the body of heavy metals via the tongue. Next to the intestines, the tongue is the most important eliminator of metals. Nowadays, when patients are suffering from mineral and metal toxins, algae preparations are also often given, since algae, through their connection with living seawater, are potent remedies against excessive physicalization, and easily absorb and store metals.

Conversely it is also true that we can use the so-called 'heavy metals' or metal compounds as outstanding homeopathic remedies if problems occur in the organism in food digestion. We mentioned already that food substances taken in from the outer world have to be transformed, so that they lose their 'foreign' character and do not become poison to the organism. The alien substance ingested has to be 'humanized' right into its molecular structures so that each person possesses his own protein. But if 'logical' metabolic stages go haywire, substances remain a part of the alien world, and external laws continue to dominate in them. These substances then irritate the organism — leading for instance to allergies — or they end up in the wrong place, or overstrain excretory organs like the kidneys or the liver. If fat or protein degenerate in outer nature we speak of them going rancid, or bad. If this natural tendency external to the human element continues in the organism it leads to various irritations in the stomach and intestinal tract, to deposits and halitosis. In such a case homeopathic metals can help the protein and the fats to be better incorporated into the organism. Arsenic combats the tendency to rancidness, and the halitosis caused by deposited fats. If intestinal problems are triggered by decaying protein, giving rise to flatulence and cramps, then copper (cuprum) in homeopathic dosage, for instance cuprum 6×, is the best remedy. This also has an anti-parasitic effect, for instance against certain 'foreign residents' in the intestines such as worms and other parasites.

Medical supplies to take on your travels, especially to the tropics, should at the very least include Arsenicum 6× and Okoubaka 2×. Besides having a detoxifying effect they also stabilize the circulation and prevent energy loss through dehydration. Arsenicum is especially helpful if a person can't for instance keep anything down, such as essential medicines. In addition if you avoid ice-cold drinks, this will greatly reduce the risk of an 'abdominal cold'.

Plant poisoning

In cases of plant poisoning, symptoms are not only nausea and intestinal cramps, as with mineral toxins, but strong effects throughout the organism with diarrhoea, cramp and dimmed consciousness.

Whilst the minerals and metals attack our life organization, plant poisonings go further, into soul structure. The whole life of feeling and sense perceptions begin to change. Whereas in the case of mineral poisonings we have to fortify the life forces as such by administering the life-bearers, protein and oils, with their connection to light and warmth, in the case of plant poisonings we must resort to substances that break down and eliminate alien life, helping to stabilize organs and re-incorporate soul life in the right way.

In nature we find such a substance in trees whose bark has a high tannin content. This substance is also used to remove protein from animal skins, and to preserve them, a process we call 'tanning'. Wherever inner and outer decay and disintegration is occurring, or something is rotting, or alien life is threatening to gain the upper hand, we work with oak bark or willow. These barks are anti-inflammatory, antiseptic and disinfecting. So the substances obtained from these are good in the case of diarrhoea and ulcers. A substance in willow bark is also of course synthetically manufactured today, and is known worldwide as salicylic acid or 'aspirin'.

Obviously there are numerous other plants containing tannin, such as sage, tormentil or bilberry, that all have their merits; but here we must highlight trees whose peripheral 'skins' especially absorb influences from the surrounding cosmos. We can mention in passing that, besides tannin, oak bark contains much calcium, which is very helpful for treating allergic disorders.

In recent years and decades, though, other potent tree barks have come into use as remedies, such Lapacho bark from South America and the bark of the West African Okoubaka tree, both of which, in a similar way to oak and willow, can be used in homeopathic dosage against food and bacterial poisoning of any kind, nicotine damage in the metabolic region, or after lengthy periods of taking antibiotics, here often in combination with Nux Vomica.

In this connection let us mention the detoxifying effects of coffee and tea, which, besides caffeine also contain tannic acid and therefore assist the digestion. Every intake of food is basically already the beginning of a gentle poisoning, so that we actually never know whether we can tolerate the food we have eaten however good it is. Coffee should, as far as possible, be drunk after the meal, tea during the meal — as is often anyway the case.

In cases of vomiting, caused by eating food that has gone bad, we recommend a strong cup of black tea (ice cold!), one spoonful at a time, in order to detoxify oneself, calm the stomach and stabilize the circulation. In many cases the same can be done with ice-cold coca-cola and pretzel sticks to stimulate the circulation, the salt replenishing lost minerals.

Animal poisoning

The dangerous effects of animal poisons are not usually found in the stomach and intestinal tract, since they are protein substances digested through metabolism and therefore rendered harmless. We can therefore actually drink or eat snake poison without coming to any harm!

They only become injurious and a risk to life when they enter the blood stream directly. The concentrated soul quality expressed in animal poison seeks to attack and dissolve our spiritual-bodily organization. One could also say the attack is directed against our ego bearer, the blood. Bodily and soul symptoms of complete and rapid dissolution after certain snakebites, for instance, are devastating. On the other hand, snake poisons in homeopathic dose are wonderful remedies for many virulent illnesses through to cancer. We owe to physiologist, Konstantin Hering, a pupil of Samuel Hahnemann, one of the most potent homeopathic remedies using snake poison, Lachesis muta from the highly poisonous pit viper. This is used, among other things against life-threatening inflammations.

We can have an almost homeopathic experience of the positive effect of animal poisons when we are stung by insects such as ants, bees or wasps. If we don't suffer from allergic reactions to any of these insect poisons, we feel a warming of our organism, manifesting as inflammation, which exerts a healing effect on 'cold' complaints such as rheumatism, gout and arthritis – disorders that threaten to over-mineralize our capacities for limb movement. In this context let us also mention the stinging nettle, which even grows in places where people have left piles of rubbish and actually has an enlivening effect on the soil. It does the same in the human organism when too much 'rubbish' has piled up, with a need for purification or purging of the blood. Rudolf Steiner, the founder of biodynamic agriculture, told farmers that stinging nettles, with their iron and sulphur content – which are needed in our blood for detoxification – are a real 'blessing' and do not deserve to be left to grow either unnoticed or scorned. If we only recognized their true value in the landscape and in medicine, he said, we would plant them 'close to our hearts'.

Where animal poisoning is concerned – and this applies for instance also to the transmission of rabies – the blood must be helped by direct injection of an antidote to neutralize destructive forces. In the case of an allergic disposition toward certain insect bites, the same toxin can be injected in homeopathic doses to gradually overcome the allergy.

Poisoning and detoxifying in a healthy organism are two activities, like breakdown and synthesis, that occur naturally in us. Everything is poisonous, our wise ancestors used to say, it is only the dosage that determines whether something has a curative or destructive effect. Even the healthiest food, enjoyed to excess, can so overtax a person that it eventually makes him ill. In this connection we can as well ask ourselves whether everyone actually ought to drink two or three litres of liquid a day, or whether we might do better to attend carefully to our 'inner physician', that is our instincts?

In order to detoxify itself, the organism has diverse possibilities of elimination, whose end products can tell us a lot about the organism's inner activity. The next chapter is concerned with this.

The Importance of Secretion and Elimination

Study of the contents of elimination during illness has long played an important part in diagnosing misplaced organic and also psychological functions occurring invisibly within the organism. Diverse illustrations depict the medieval physician carrying out a 'urine inspection' with a urine glass in his hand, and thus ascertaining the patient's state of health. Tradition tells us that the urine produced at the first 'crow of the cock' and collected in a transparent vessel, was protected from warmth and sunlight and brought to the 'urine inspector' who examined it in its fresh state and again two hours later for its density, colour, smell, taste and sediment. So-called 'urine cards' actually circulated among the people for this purpose in the Middle Ages. After determining the colour, smell, taste and the sediment of one's own urine, one could consult these cards for self-diagnosis.[62]

An illustration from the sixteenth century shows two doctors, one of whom, the 'genuine' doctor, is carrying out the traditional urine test, while the other, the 'quack', is instead 'studying' his purse. A theme that still remains topical today!

When speaking of 'secretion' below, this concept should not be understood too narrowly. Firstly, of course, it concerns 'classic' excretion such as sweat, urine and the contents of the intestine, that are essential to the detoxifying of the organism. But all the glands continually exude secretions and enzymes in an inward direction. So besides the outward elimination, we can also speak of an inward-directed cleansing. Under 'elimination' we can include everything that the organism either inwardly secretes or outwardly eliminates. The term therefore covers everything tending to become lifeless but which either has been or still will be used by the organism. All exudations, even hair and nails, can be included in the diagnosis of illnesses. The placenta too is an eliminated product that the maternal organism and newborn child do not need any more, though it has played a vital role.

Everything excreted or eliminated was originally subject to soul-spiritual activities, and still bears the 'finger prints' of forces previously functioning invisibly. Interestingly enough, these end products used to be called 'mummies' in German, to describe an entity previously filled with psychological and spiritual life.

In this respect we can ask—even if this is a shocking idea—whether the brain, too, isn't to be seen as a non-living, permanently 'decaying' mass, as a 'dead mirror', an inwardly structured 'sedimentary' organ, albeit of the noblest kind, and exuded from cosmic thought-forces; and thus not a lump of protein that only 'secretes' thinking but one through which our own thinking activity comes to consciousness in us.

This most highly refined substance, 'secreted' upward, corresponds in a polar sense to the contents of the intestines in our lower system, whose superfluous waste products are outwardly excreted. Thus we can regard the brain as the opposite pole to this, and its 'higher intestinal content'—without the protective and enclosing wall of the intestine—as the noblest and most refined 'mummy'. Acquainting ourselves with this—admittedly unusual—way of seeing things, we can come to understand pathological connections between the brain and the intestines, and recognize the therapeutic possibility where brain problems are concerned—even if only a migraine—of achieving significant things through treatment of the intestine and digestion. We have to embark on such treatment very low down in the metabolism to achieve success with this 'end product', whether manifesting as migraine, neuritis, inflammations of the sense organs, or even brain tumours. Conversely, as already mentioned elsewhere, brain-like structures are independently active in the intestinal region, and process soul impressions. We can recognize something of this if we say that we need to 'digest' something, or speak, quite intuitively, of a 'butterfly stomach'.

In our physical body, therefore, we have at one and the same time a polarity of synthesis and breakdown. We find synthesis in the lower system especially in the forming of substance and in life forces, and breakdown in the upper system, which is the bearer of our conscious soul and spiritual organization. Where spirit appears, material must depart and relinquish its vitality, something we can witness very clearly in our sense organs, nerves and the brain. Thus processes of breakdown and elimination are linked to our consciously or unconsciously functioning soul life. We are aware in our ordinary everyday experience of the connection between soul activity and spontaneous exudations — sweating in fear, the need to urinate, diarrhoea when upset, or increased production of saliva at the mere thought of our favourite dish.

In traditional medicine and also in modern holistic medicine, inspection of excreted products such as sweat, urine and faeces are an important diagnostic tool. One just needs to know how to interpret them: quantitative analysis is not enough; the qualitative aspect is also vital, with observations of colour, form and locus of elimination.

Perspiration as the 'mummy' of life processes

The thinnest, clearest human secretion is sweat. For exact diagnosis it is important to determine where it occurs in the body and at what time: during the day or night.

Besides warm and cold sweat, we know of sweat that smells bad or has no smell at all, but also sweat that can be partly exuded from particular areas of the body: the feet, the back of the head — as for instance in rickets — or other parts. Perspiration is healthy after psychological or bodily effort, but there is also sweating that comes out of the blue, as for instance during the menopause. Night sweating, when we are neither at work nor consciously active otherwise, indicates a possible lung problem. But it is quite normal that we sweat a little when we 'enter' our body again in the morning after sleep.

In general we can say that the water in us, the sweat, represents our life forces, and we therefore speak of a 'depleting sweat' or also of a 'cold' or 'deadly' sweat.

In holistic medicine, there are many remedies that can inhibit or promote sweating. Sage and deadly nightshade (belladonna) counteract excessive perspiration, and baths of hay flowers, and elder blossom tea, help healthy perspiration to develop. If the skin eliminates properly, one doesn't need to worry about the inner organs. In anthroposophic hay-fever therapy, after certain remedies have been injected, sweat baths are given to encourage the body to 'sweat' in the right places and not in the wrong places – such as through the nose or other mucous membranes.

The manner and the site of perspiration, but of course also the reverse, dryness of the skin and the mucous membrane, tell us something about the psychological and bodily condition of the person concerned. There is for instance an interesting remark of Rudolf Steiner's that I can thoroughly confirm in my practice: 'If someone's thinking is trying to destroy their head then the organism helps itself through foot perspiration.'[63]

In psychological disorders, one can sometimes notice bad-smelling foot perspiration a few days before the symptoms grow drastically worse – a sign that the organism is starting to defend itself against inner toxicity, which it can do not only by sweating, but also by suppuration. Rubber boots are therefore not the only cause of stinking feet!

The interesting thing about these symptoms is the relationship between the psyche and sweat formation, on the one hand, and on the other also the physical locus: though head and feet are far away from each other, there is an inner connection between them. This helps explain why footbaths are useful when illnesses such as migraine, sinus inflammation or flu-like infections occur.

Urine as the 'mummy' of soul activity

Successes with auto-urine therapy have in recent years raised awareness of the elimination products of the kidneys and bladder. Indeed, in chronic illnesses there have sometimes been almost miraculous cures through intake of one's own urine, or external applications and injections of it. What is so special about this particular fluid?

Salty sweat, mostly without any smell or colour, is the

expression of life forces, whereas mineral-rich and coloured urine is the expression of the soul activity that engages particularly in the kidneys and bladder.

At every moment our various soul activities influence the formation of urine in the kidneys. We are unaware of this as the bladder collects the urine unnoticed until it makes itself felt in 'bladder pressure'.

In studies undertaken in 1928, the quantities of nitrogen in the urine (uric acid and urea) were measured in association with objective, abstract thinking and with simultaneous cold or warm applications — cold to the head and warmth to the feet. It was found that more of the breakdown product of uric acid forms during abstract, cool reflection. We can sum up as follows the study findings:

1. The elimination of uric acid and urea is markedly affected by objective thinking in that more uric acid is produced but less urea.
2. Application of cold increases this influence.
3. Through simultaneous use of warmth the changes described are characteristically changed into their opposites.
4. It was demonstrated that cold and warmth applications do not lead to this effect on their own, and that it can therefore only be ascribed to thinking activity.[64]

This can make clear why coffee, which stimulates production of uric acid in the organism, encourages thought activity and makes a person awake in his neuro-sensory system.

Urine represents the 'chemical mummy' of the organism's previous soul activities, with their tremendous number of products such as uric acid, hormones, minerals and oxalates, which can all tell us something of significance. And if the inner organs are too weak really to transform protein and sugar properly so that these are eliminated in large quantities in the kidneys, then the organ will be strained or damaged. If breakdown activity progresses too rapidly, and if the blood even starts flaring during fever or inflammation, then various kinds of cloudiness appear in the urine. Fleecy urine has sometimes been compared to a 'stormy summer day' when dark clouds appear. This can indicate incipient fever or inflammation. Conversely, breakdown products can be held back in the organism so that the urine appears bright yellow to white, reminiscent of a bright summer's day. If this lasts for a long time then we have to ask why the organism is holding back its waste products and what the consequences of this will be for a person's health. 'If you have dark urine then you have a tendency in some way to inflammation in the body. If you have a very light coloured urine a tendency to tumours.'[65]

In traditional oriental medicine, inspection of the urine is still taken very seriously and its quality is tested by stirring with a little stick, naturally occurring foam being included in diagnosis. In the case of excessive foaming the catabolic soul forces of the kidneys are too active.

Similar to autologous blood treatments, we can understand the action of auto-urine therapy in these terms: everything dispelled from the organism is subject to other laws, namely earthly ones. If this is returned to the organism, the latter must again defend itself against an alien element, yet one which was previously part of it. This is roughly like trying to resolve repressed problems that return to us again in a changed form, so that we grow all the stronger in the process. In the case of urine the forces of nitrogen, especially, have a relationship to the soul functions, and call upon the organism to reintegrate inner, bodily forces. In modern orthodox medicine, too, urine-based ointments are increasingly being used.

Intestinal excretion as the 'mummy' of I activity

By now I hope it has become clear that the whole human being is continually involved in elimination processes, and that visible and measurable excretions represent only the sum of these functions. In this sense the intestinal organism is a kind of 'drainage apparatus' for end products. By means of it, solid matter, that is, the remains of solid food, is eliminated, signifying the greatest exertion of the whole organism. The ferments of bile, stomach juice and pancreas have to be extremely aggressive in order to overcome the outer world and to rid food of its own nature. This activity is achieved by means of the human ego organization, which, in a wonderful alchemical process, can transform natural substance into our own human substance.

Depending on the strength or weakness of this ego, the process of transformation will be complete or products will only be partially worked through. This can just as well be caused by bodily weaknesses as by psychological ones. We notice it when, for instance, we suddenly cannot tolerate a particular food, though other people had no trouble in the night with, say, the shrimp cocktail eaten in the evening.

The ego has to balance 'too fast' against 'too slow' in the digestive process. It is a known fact that we can react with diarrhoea if strongly psychologically affected by something; or can get constipation if we are psychologically blocked or, at the beginning of a journey, have to adapt to new rhythms and the new surroundings. There are of course innumerable variations on this theme.

The case history of a patient whom Rudolf Steiner attended in an advisory capacity shows us this context once again, and the therapeutic response to it.

> The regulated digestive activity of a human being is eminently dependent on normal ego organization. Enduring constipation shows a failure of this ego organization. From this disturbed digestive activity follow migraine-type conditions and the vomiting that she is suffering from.[66]

Among other things, copper ointment was placed on the kidney area as a compress, to exert a strengthening, warming effect on the ego organization. It is interesting that this regulating of the kidney region assuaged both mental/emotional *and* organic complaints, namely fear and pounding of the heart, that occurred especially on waking up. As we saw in the chapter on the kidneys, both of these have to do with the kidneys.

Thus, however unpleasant it might be for the nose and the

eyes, it is clear that inspection of solid excreta, their colour, form and undigested parts, form part of a full diagnosis.

If we assume that our eliminations, these 'mummies', are the material end products of soul-spiritual forces, we can, with an eye to greater fixity and constancy in the animal kingdom, learn what to do in the case of human illnesses. A human being creates individual variations in the excreta between the extremes of 'too fluid' and 'too hard,' whilst in the animal kingdom, for instance, with a cow or a hippopotamus, a hare or a horse, the eliminations are 'fixed and constant' except in the case of serious disorders. Thus sloppy cow dung gives expression to the phlegmatic nature of this animal, whereas the round, drier 'orbs' and 'pellets' of the horse and rabbit demonstrate their joy in movement and sanguine hopping. This is what people of old understood by a 'mummy': something that revealed the secret of a creature's whole inner being in the excreta.

If human beings suffer, like the metabolically-oriented cow, from stools that are too liquid and sloppy, this is a sign that the formative neurosensory forces are engaging too weakly, and then we should use specific remedies or a diet of roots, with their specific relationship to nerve organization.

Conversely if the dissolving forces of digestion are too weak, then little round, dry excrement—called 'sheep dung' in homeopathy—arise. Here the nerve-sense forces predominate, which needs balancing by strengthening the metabolic forces with blossom and fruit and certain sulphurous medicaments.

So what the ancients called a 'mummy' is worth another look, from a new, more comprehensive point of view. After all, ancient Chinese physicians maintained that 'good luck begins with digestion!'

Nutrition and Healing

A good doctor looks first of all for the cause of the illness, and if he has found it he tries first to find a cure for it through food. If that doesn't solve it he prescribes medicine.[67]

Sun Ssemiao

Our daily nourishment relates essentially to maintaining the health of body and soul and, in the case of illness, involves certain dietary principles. Today, traditional ways of preparing food and a proper culture of eating are giving way increasingly to 'fast food' habits; from an early age healthy instincts are manipulated and seriously affected by chemical food additives such as synthetic aromas, synthetic sugars and taste enhancers. We now need to reclaim our predecessors' art of cooking and understanding of food through a deeper understanding of their meaning for our bodily and psychological well-being. But to eat only with our 'heads' is no solution either, as to be nourished is really a matter of wanting to eat and enjoying doing so.

Founded on a long tradition and on joy in the physical world, eating in Asia is very much a spiritual-social concern, and is discussed with passion. Even the poorest, simplest people have a deep understanding of its value for health. In the widespread Chinese understanding of food, life does not lie in the lap of the gods but in the hands of a good cook, and a considerable part of the joy in life has to do with eating, which in Asia is not meant quantitatively. A close connection has always been seen there between wisdom and good food, and it is an infallible sign of a wise man that one eats well with him. A legend even tells us that the famous Confucius separated from his wife because her art of cooking no longer satisfied his taste.

In the oriental perspective there is no significant difference, even today, between a dish of food and medical remedies. For anything that is in any way good for the body is at the same time salutary for it.

In Indian Ayurvedic medicine, we are told that medicine suitable for human beings has to be founded on three therapeutic pillars:

1. on a change of consciousness;
2. on a change of food in the sense of a certain diet consisting of refraining from particular dishes; and,
3. last and least, on the use of remedies that are, ultimately, only the 'spice in the soup'.

Europeans may smile a little when they read, in a book about the East-Asian way of life, that the real reason the Chinese have not developed a rigorous botany and zoology is because a Chinese scholar cannot, for instance, look at a fish coldly and heartlessly, without immediately wondering how it might taste.

As far as our daily food is concerned, we are always somewhere on the scale between over-valuing food—which can sometimes turn into fanaticism—or a criminal neglect and ignorance, because we underestimate the actual significance of a healthy diet—something we can see ever more clearly today in the catastrophic state of physical and mental health of the population. We need think here only of the increase in obesity and eating disorders even in young people. Rudolf Steiner's answer to a question from the pioneer of biodynamic agriculture, Ehrenfried Pfeiffer, as to why people knew so much yet did nothing about it, was:

> This is a problem of nutrition. The nature of food today doesn't give people the strength to make the spiritual manifest in the physical. The bridge can no longer be created between thinking, will and action. Food plants no longer contain the forces people need.[68]

Steiner's view of real nutrition and nourishment, of which—according to Pfeiffer—he spoke every time they met, talking about the secrets of the digestion and the re-enlivening of plant life, is not unlike that of an ancient philosopher: that spiritual, moral development must be supported by the right diet.

But since the spirit has become so soulless and abstract, no one today would give serious attention to a philosopher who—like Nietzsche—embarked on the subject of eating and drinking:

> Philosophy is nothing other than the instinct for a personal diet. All philosophising has hitherto been a matter not of 'truth' but of something else, let us say health, the future, growth, power and life.[69]

Even if there are no hard and fast rules in this vital realm, we can certainly ask what we could imagine a 'healthy' meal to be?

A few thoughts on this:

Any kind of food that gives a person the least trouble and stimulates and supports his physical and spiritual activities to the greatest extent can be described as 'good'. For this we must develop our individual taste so that what we eat and drink corresponds with our character and our constitution. In the course of our lives our habits often actually change if there is a decisive change in our consciousness. But in general it is nevertheless good from earliest childhood to acquire, through imitation, the right foundations that then become good habits.

Through a deepened knowledge of processes and the actual 'spirit' in matter, as these unfold in us through the preparation of food and eating, a freely individual relationship to food can come about which, as already emphasized, is determined by our constitution, our temperament and particular life circumstances. We will illustrate this with a few examples, and also describe how this knowledge can help us either to use certain foods or avoid them.

Of the substances that we ingest daily, foodstuffs are polar opposite to medicines, the 'spiritualized' substances. Between these two extremes are the spices and stimulants, which interestingly have a connection with both 'sides'. For our digestion the spices are 'remedies' that help to transform properly what we take into our organism.

Based on the threefold nature of human beings and plants, and their dynamic reverse correspondences (neuro-sensory system = root, rhythmic system = leaves, and metabolic system = blossoms and fruit), as well as the polarity of cosmos and earth, buoyancy and heaviness, as these appear clearly in blossoms and root, we acquire a first qualitative orientation in the choice of foods, either to consolidate our earthly bodily nature by eating roots or by filling ourselves with cosmic light and warmth forces by eating blossoms and fruit. Thus the primary food for an infant is carrots, and if they have rickets then a bath therapy using calamus root, for instance, is indicated. None of this of course should be understood in a one-sided way, for the cosmic principle must always also be present in the form of fruit. But this aspect becomes interesting when we consider that carrots and especially beetroot show an interesting 'shift' towards being a medicinal food: what normally takes place in the blossom — colour, scent, sweetness — makes its way into the root in carrots and especially beetroot; in which case they actually become a 'blossom root'. It is not for nothing that beetroot with its colouring (flavonoids) is recommended as a cancer diet. This 'blossom' principle, or rather the developing of light and warmth, can also occur in substances that become sulphurous such as radish and horseradish, both of which ' clear' the head and give a spur to thinking.

In this polarity, of light and warmth/cold and moisture, are all the substances that we obtain either from fruit, blossoms or roots. In some schools of nutrition people speak of the parts of a plant that are either 'above' or 'below' the earth. There is a qualitative difference whether we take sweetness as honey from the realm of blossoms or from the root as sugar-beet, and then often mineralize it further. The latter stimulates head forces, thus also causing tooth decay, whilst honey, as 'the milk of old age' has an anti-sclerotic effect and connects us more with the cosmos again. It is obvious that honey is not a food for infants, and small children given too much honey by well-meaning mothers can sometimes get severe diarrhoea and rickets.

In ancient Greek medicine people thought they could protect themselves against injurious earth forces by honey from within and oil from without, both of which are formed from cosmic light and warmth forces. Here too there is a difference whether we take olive oil whose fruits have to ripen for years in the southern warmth, or peanut oil from a leguminous plant that forms a great deal of nitrogen and therefore also subtly toxic substances, and grows in the dark, damp soil. The same applies to the potato. As a tiller plant containing a tremendous amount of starch, whose digestion has to be completed in the head, and is very heavy fare for migraine sufferers, it is not an ideal food. This is by no means to suggest it shouldn't be consumed! But one needs to be well acquainted with it since, in contrast to grains, for instance, potatoes only properly form starch in cold conditions, and under the influence of light produce toxins (solanine) — familiar as green areas. Like most nightshades, such as tomatoes, aubergines and pepper, but also maize and peanuts — all originating from the American continent — this 'heavy' food is not an ideal diet for the very sick, for instance someone with cancer, who needs light and warm substances formed in the blossoms and fruit above the earth, and also sulphur-rich cabbage. It ought to be evident that fungi, too, growing on decomposing soil and feeding on detritus, are not an ideal food, however tasty they may be.

Thus we can get our first bearings on nutrition if we understand the plants not only in terms of their constituents but also qualitatively, and relate these qualities to the human being.

Of course, points of view such as whether they are cooked or raw play their part too, and have to be judged individually according to available metabolic forces. Cooked, fermented and pickled foods are already in a sense pre-digested, whilst raw food, and all purely vegetable food, has a curative effect. This is why raw food should only be used as sole diet for a short time, that is for a week at the most, depending on its effects, and should not predominate in the diet, except in exceptional cases. Vegetarian plant food connects a person more strongly with cosmic forces, while a diet emphasizing animal protein brings us more strongly onto the earth, and 'consolidates' us more. This applies also to protein — i.e. nitrogen-rich pulses such as peas, soya, lentils and beans: these make us 'heavier', which can also be good for extremely 'airy' constitutions.

The substance that holds an ideal balance and is neither a

purely plant substance nor a purely animal substance is milk, with its diverse products.

The constituents of food are chiefly protein, fats, carbohydrates and trace elements, which — apart from the minerals — have their specific way of degenerating: proteins decay, fats go rancid, carbohydrates ferment. As foreign protein contains the most life forces, an excess of animal protein can lead to mild poisoning, and in a poisonous form can even be life-threatening. In cases of severe illness, a tendency to sclerosis, and during epidemics, we should strictly limit our consumption of protein!

Natural fats like butter and olive oil are ideal substances for the blood stream and the heart and help build up the brain and nervous system, which of course consists mainly of fatty substances. The fats ingested are not, like proteins, destroyed immediately in the intestines, but enter the bloodstream, and are only fully digested when they reach our centre, the lungs. Hence their 'sympathy' for the life of feeling and for warmth. Cholesterol is raised on the one hand by stress and on the other by degenerated fats, the oxicholesterols. Everything no longer alive remains as it is and becomes a poison, which is why ultra-heat-treated milk represents a problem.

So we must give our children a living diet from an early age so that their organs, the brain for instance, can develop properly. Research has shown that mother's milk is an ideal food for healthy development of the brain, which is increasingly under attack from a poor diet and certain additives.[70]

In this connection there is an interesting remark of Rudolf Steiner's from the year 1904, about healthy development of the brain in growing children:

> If someone desires to school their thinking they will need to have above all a well-developed, healthy brain. But today(!) parents seldom provide their children with this healthily structured foundation, so remedial help is needed to strengthen it. And here above all, hazel nuts supply the necessary substance[*] for brain development.[71]

---

[*] Phosphor-rich fats, i.e. lecithins.

Given global genetic manipulation and modification, and impoverishment of food quality, we should be very concerned about humanity's healthy development of mind and body. The greatest efforts must be made in education, agriculture and medicine to ensure that food is not robbed of its cosmic value — unless of course countries desire ever-increasing numbers of brainless subjects no longer interested in healthy critical judgement. Therefore let us urgently point to dangers of something which, with the best will in the world, and often through lack of time, unwittingly slips into our lives through the back door like the devil. It may seem we are doing something beneficial, and saving time, and that there's no need for concern since foods still retain their vitamins. I'm thinking of the microwave. Is something still alive just because its substances still exist? Is a person still alive if I beat him to death, open his body up, and say that he is still alive since all his organs are still there?

Apart from the fact that Swiss researchers have discovered that microwaved food leads to altered proteins in the blood, we must above all consider the effect on food of high-energy rays. As background here we need to know that electricity is never friendly to life, but instead is attracted to organs that tend to be less organically alive, for instance our brain and neural system, or the root of a plant. During the agricultural course to found biodynamic agriculture, in Koberwitz near Breslau, Rudolf Steiner was asked by a farmer whether it was a good thing to conserve animal food with electric current. After speaking in detail about the nature of electricity — which is already creating enough problems today in our global internet age — Rudolf Steiner described what outcome such animal husbandry methods would have, and made clear that these effects also apply fully to the human being:

> So if we use electricity to electrify fodder, we produce food that is eventually bound to lead to sclerotic degeneration in the animals we eat. It is a slow process — which we won't at first notice. All we shall notice at first is that these animals die earlier than they should. We won't realize that the cause of it is electricity but we

will attribute it to all sorts of other things. [...] but living creatures
will gradually become nervous, fidgety and sclerotic.[72]

Today there are studies that show that water and animal feed
treated with microwaves leads in a few generations to depletion
of reproductive forces in both plants and rats.

We saw earlier how a soul-spiritual 'diet' and a particular kind
of nutrition must complement one another. It is clear that great
creative human works have in the past been garnered on a
meagre diet, or even in states of hunger, while people who eat
too much and too often have to expend excessive energy on
digestion, retain scarcely any capacity for creative thinking. ('A
full stomach makes studying hard!') The mind is never so clear
as when it is fasting, and in the biographies of many saints, who
often lived only on fruit, bread and water, we find a great many
examples of the effects of an abstemious life. By limiting food,
and fasting, we can diminish the irregular effects in soul life
arising from voracity, and the forces of life can once more unfold
their original cosmic rhythms.

This knowledge was common to every great culture that
connected religious practice with certain dietary or fasting cus-
toms. This is the origin of the idea that moderation purifies the
feelings, awakens slumbering capacities, gladdens the soul,
strengthens the memory and lifts the soul's earthly burden so
that it can delight in a higher freedom.

When something as important as our nutrition is at stake, not
as an end in itself but also serving the further spiritual development of the individual and of humankind in connection with
maintaining the health of the earth, western and eastern wisdom
come together again. Our concerns here can be summed up in
these beautiful words:

> Anyone who attaches importance to his health must be moderate
> in his taste, avoid worrying, calm his desires, moderate his feel-
> ings, nurture his life forces, be sparing in his words, not think too
> highly of success or failure, hold worries and difficulties in con-
> tempt, give foolish ambition its marching orders, avoid excessive
> inclinations and dislikes, use sight and hearing with composure

and remain true to his inner diet. How can one be ill as long as one has not worn down one's courage for living and saddened one's soul? Therefore whoever seeks to strengthen his nature should only eat when he is hungry, and not fill himself with food, only drink when he is thirsty, and not enjoy too much drink. He should aim to remain a little hungry when he has eaten well, and always to feel a little replete even when hungry. To be well filled harms the lungs, and to be hungry inhibits the flow of living energy.[73]

Naturopathic and Herbal Medicine, Homeopathy, Chinese and Anthroposophic Medicine

Now the peculiar thing with the Chinese, however, is that they cannot think in concepts at all but only in pictures; but then they place themselves right inside things. And they are actually able to make all the objects created by external invention, unless it be things like steam engines or suchlike. And the way the Chinese are today [...] is only how they have become after being mistreated by Europeans for hundreds of years [...]

Now, you see, a Chinese person actually has a great interest in the environment, a great interest in the stars, and a great interest in the outer world in general.[74]

Rudolf Steiner

Readers of this book may have asked themselves why the author has repeatedly drawn attention to Chinese medicine and its vivid pictures of the organs.

Does something have any relevance still if it is over 2000 years old, and arose from a quite different cultural background, thus a totally different understanding of human beings, nature and the world? Experience shows that western-oriented scientific medicine, with its analytic-quantitative thinking, is increasingly debating the holistic views of Chinese and ancient Indian (Ayurvedic) medicine, and integrating them into medical practices and hospitals, especially to treat chronic illnesses and pain. Of especial interest here are Chinese acupuncture, certain massage techniques, and a few tried and tested medicinal herbs, some of which are now sometimes used to treat malaria. It remains to be seen whether, at the same time, the spiritual roots of this medicine—far deeper than merely technical use of needles—will be properly discussed, let alone understood.

Many Chinese words are scarcely translatable into our concepts because they have double or multiple meanings. Even the word 'doctor' — 'yi sheng' in Chinese — means, roughly, 'a life healer' and therefore has a much more comprehensive meaning than we usually understand by the word 'doctor'. Even the concept known in the West as chi, roughly 'life force', has several meanings. For the Chinese language knows no abstract concepts but always acquires its specific meanings by combining words. In Chinese chi means more or less 'air, steam, breath' or also 'gas'. Added to mi we get 'rice', which surely has to do with 'nourishing steam'. The pi chi would then be, for instance, the chi of the spleen, and means 'a bad mood'. Thus the Chinese distinguish the native chi life force, that we have inherited from our ancestors and is stored in the kidneys, from the acquired chi that someone obtains from the air, food and social relationships. To regenerate diminished chi again, the doctor, as life healer, recommends that, beside medical treatment, the patient gets a healthy diet, bodily exercise, good sleep, good friends and laughter. We see from this that the Chinese invoke not only life energy but also the kind of energy that comes from the soul. According to the Chinese view the chi in the organism has its own circulation that affects the blood. Who or what governs the cosmic life force within us? This is determined by the 'intention of thought', the yi, and is therefore a psycho-physiological force related to the rhythmic breathing process, the energy of the blood and unspoken thoughts. We could perhaps equate it with the soul force streaming into us, connecting us to the outer world in a breathing, feeling way, in which arise, among other things, feeling disturbances such as fear, paralysing our whole vitality. 'In the case of great anxiety and sudden fear, our consciousness melts away, the blood and the chi separate, and yin and yang burst apart bringing chi into confusion.'[75]

Can we still understand these ancient thoughts and pictures today and simply translate them into our philosophical or other systems of knowledge? How for instance does the Chinese five-element system — earth, water, air, metal and wood — relate to the four classic elements earth, water, air and fire that arose in Greek times and form the basis of natural medicine and the theory of temperaments? Are the Chinese 'phases of transformation' identical with 'metamorphosis'; the cosmic opposites yin and yang with 'polarity' or the life force *chi* that we mentioned, with the life forces in human beings? We see in these few examples that a lot of further reflection is needed. Although there are certain similarities and agreements between Chinese and anthroposophic medicine, the conceptual background is entirely different. We will examine this in more detail below.

One thing however connects all these holistic medical ideas: they embed people again in their universal relation to the cosmos and in nature, and, taking their departure from a certain picture of the human being, try to make them 'whole' again when they have lapsed from wholeness through illness. All healing measures, after all, ultimately seek to treat both the physical body and the soul and — especially in anthroposophic medicine — besides medicinal and artistic therapies to help the individual human ego gain spiritual knowledge about himself and the world. In this case it is important, whether this is expressly stated or not, to create and express a third principle between the polarities, or the yin and yang: something new, a mediating principle that corresponds physiologically to rhythm and seeks to harmonize life forces and soul forces, and bring about a true meeting between doctor and patient. Through this can arise something that, for once here, I would like to call the 'Christian' element.

During a visit to an Ayurvedic clinic in Sri Lanka a few years ago, I became aware how one-sided our western medicine often is in its procedures when it equates 'healing' with the disappearance of symptoms. On leaving, I asked the senior doctor whether he could briefly outline for me the fundamentals of Ayurvedic medicine. He explained to me that there were actually three pillars sustaining it:

1. changing of consciousness, thus one's outlook on life;
2. changing nutrition, in the sense of adopting a certain diet; and
3. finding the right natural remedy.

All three were necessary, he said, for a full cure.

I would like at this point to try to present an outline of the basic principles of natural therapeutics, homeopathy, Chinese and anthroposophic medicine. Due to brevity there is a risk that this will be very superficial.

Natural and herbal medicine is familiar to most of us from our childhood. Traditional medicinal plants such as melissa, camomile, stinging nettle and wormwood often help us when we are mildly indisposed. Fasting cures, certain diets, compresses, purgatives, especially, too, the use of the four elements that we find as much in nature as in the human being, are usually accepted without question because they have proved themselves empirically for thousands of years. There is probably no need here to stress the benefits of medicinal clay, foot baths, oxygen, carbon dioxide and the many different kinds of heat application.

'The doctor helps while nature heals' is of course a well-known motto of natural medicine. The basic idea is that the organism's processes of elimination must be stimulated to activate forces of self-healing. There must be a secret bond between nature and the organism for certain remedies to be able to act on particular organs at all.

Although homeopathy also uses remedies from the three kingdoms of nature, it does so in a way that goes far beyond herbal medicine's substances, potentizing and dynamizing them, thus, as Hahnemann, the founder of homeopathy, says, 'spiritualizing' them. One could also say: the energy within matter is released, gradually discarding its material dress through the process of potentization. To understand this we can imagine an iron magnet with whose help, by gradually distancing an iron object from it, we can capture and increasingly concentrate the magnetic energy itself in a particular medium.

Medicine testing on human beings, during which mild symptoms of poisoning were produced by certain substances, has given homeopathy a tremendously differentiated store of symptoms or 'pictures'. When ill, we can then resort to the corresponding remedy in homeopathic potency. In ideal cases the toxicity or remedy picture's chief symptoms correspond precisely to the illness. A normal bee or wasp sting is already a natural medicine test in a healthy person. It produces certain symptoms, which, when they occur in an illness, are treated with potentized bee poison (apis). Thus the basic homeopathic principle is 'Like cures like', in that the organs' self-healing forces are activated. The physician can learn from homeopathy how abstract descriptions of illnesses can be transformed into a 'physiognomy of illness' and thus become individualized. One stomach ulcer or migraine is then no longer like another. What does a doctor usually resort to when a patient attends with a migraine other than an analgesic? But if we want to treat a cause as such we shall have to consider all the migraine symptoms to find the right remedy: the cause of the pain, where it is located, the connection with other symptoms, whether this is menstruation, constipation, vomiting, hemiplegic headache on the right or left side, psychological outlook and constitution. In cases of acute symptoms, of course, there is no objection to painkillers, but it is nevertheless doubtful whether *effect is the same as effectiveness*. Homeopathy turns an anonymous person into a specific individual with symptoms relating to his particular illness.

I myself owe much to homeopathy—not only because of its magnificent store of remedies and keen interest in nature, but also for the absolute necessity of observing much more precisely, and coming to a better understanding of interconnections by starting from the *illness pictures*. I still well remember my first success in homeopathy with the snake poison lachesis—prescribed in the case of severe inflammation (phlegmon) that appeared bluish-red—when everybody thought I should be prescribing antibiotics. That would have been correct no doubt from a bacterial perspective, but bluish-red is one of the cardinal symptoms of lachesis, and the result was amazing! Nevertheless, it has always seemed to me that something is lacking in homeopathy. Just learning the symptoms by heart never satisfied me, for I wanted to know why and how an 'immaterial' remedy worked at all, and what the connection is between the three natural kingdoms and the human being. I found the answer in anthroposophic medicine. I also had to ask myself whether, because of environmental stress, impoverished nutrition, electrical influences, psychological strains and lack of movement, we should find additional methods of healing, since, at least in our modern civilization, we can discern increasing rigidity in people's soul and life forces. Healing must therefore acquire a different dimension from over 200 years ago. Surely far greater efforts are needed by both the doctor and the patient to elicit greater inner and outer activity, and evade the 'mental cinema'?

Let us turn now to the basic principles of traditional Chinese medicine (TCM) of which only acupuncture is really known to most westerners.

Whilst our scientific worldview is based mainly on the microscope and analysis of substance, used to understand causal chains, the worldview of Chinese medicine is founded on the 'macroscopic' picture: on an understanding of nature, the universe and their underlying laws. Not until he possesses this understanding is the Chinese doctor in a position to understand illness as disharmony and deviation from these universal laws, and to recreate harmony by balancing the opposites of yin and yang again. If we learn to follow again universal laws (the Tao or 'the way'), in the Chinese view we are rewarded by long life, happiness and health.

The Chinese yin and yang theory is a dialectic logic that explains momentary connections, patterns and changes in the organism, but not in the western perspective of largely past causation. In the Chinese view, truth is immanent in things, while from our point of view phenomena have something underlying them, a transcendence. For the Chinese physician, knowledge is the wish to understand the mutual, reciprocal connections or the pattern within the dynamic of yin and yang. Signs and symptoms in the body and the soul are combined to gain an understanding of what is happening.

> Western medicine has to look behind a veil of symptoms to discover a pathological mechanism, and therefore needs a theory that goes beyond the doctor-patient situation, and depends on additional knowledge. The Chinese doctor on the other hand seldom looks further than the patient himself; theory is necessary only for guiding his perception.[76]

In western medicine diagnosis means an exact, quantifiable description of as narrow a context as possible in order to discover a single cause. A Chinese doctor looks at the overall physiological and psychological process. All available information, including symptoms and other characteristics, are linked together, until what the Chinese call the 'pattern of disharmony' becomes apparent. This pattern describes the situation of 'imbalance' in each patient. This kind of diagnosis does not, therefore, lead to a specific, isolated illness or to particular causes, but gives an almost poetic — but very useful — description of the pathological condition. What interests the Chinese doctor is more the relationship of X to Y at a given time, rather than which X causes the Y. The whole configuration of the pattern of disharmony represents the framework for treatment, which, drawing on a wide range of applications, endeavours to restore harmony in the individual.

Using the 'theory' of yin and yang in relation to specific bodily processes and organs, the Chinese doctor is able to recognize the pattern of the disharmony and treat it appropriately.

In my personal experience of Chinese doctors I can say that a patient's past or biographical situation, with psychological problems, plays no significant part in treatment. It is hard to imagine such a thing as a 'biography counsellor' in China. No consideration is given to the ego aspect. Sitting down before a Chinese doctor, and about to tell him about oneself, you find that you must, instead, show him your tongue. Before you know it, his hand is taking your pulse, and after a few questions you find yourself lying down with needles sticking into you. This is not to say anything, of course, against the value of this kind of medicine, but only to question whether it can rescue our western mentality from its modern medical malaise.

But it is striking to see how such a physician acts in direct response to momentary symptoms, with regard only to how the individual relates to and depends existentially on the whole world organism. I think this was also Rudolf Steiner's view when he spoke of the Chinese being able to place themselves right into phenomena without applying any 'transcendent' theories, which only cloud unprejudiced appraisal. In fact, homeopathy and anthroposophic medicine show something similar in this respect, when they speak of the 'physiognomy of the illness' or of the 'ideal organization of illness'; yet they do also look for 'underlying' realities. More than homeopathy, as a purely empirical medicine, Chinese and anthroposophic medicine regard their practice in terms of a pattern of order in which 'separate parts' inherent in the world and the human being have an inner connection that must be discovered in superordinate ideas.

As an example I will describe the Chinese view of the liver as a function and not only as an anatomical organ. Chinese medicine defines an organ in the first place not in terms of its physical structure but the functions associated with it.

East is the cardinal point assigned to the liver, the place on earth where the sun's life energy first displays itself. As the organ of vitality it is associated with the spring and the colour green, that is, with growing and flourishing. Its time of day is the morning when a person possesses as yet unused energies from the night. Its element is 'wood', representing both sappiness and stability. The corresponding planet is Jupiter, as we have also seen in anthroposophic medicine. Of nature's elements it is the wind that brings water into movement, and the corresponding taste is 'sour', as a water-regulating and enlivening force. The liver 'opens' to the world in eyes, the tears—i.e. the watery element—and sight. The ligaments, i.e. the beginnings of the muscles, including the dynamics of the muscles, are influenced by liver function. The corresponding soul mood is anger, violent agitation.

How do the polarities of yin and yang play into this? All the organs and their functions are subject to them, manifesting in warmth, cold, synthesis and breakdown, abundance or lack of energy.

'Yin' in its original meaning means more or less 'the shadowy side of a hill' and is associated with concepts like feminine, moon, dark, night, water, negative *chi* energy or also receptivity.

'Yang' means 'the side of the hill the sun is shining on' and its quality is masculine, sun, light, positive *chi* energy, heaven, fire and creative force.

In western-oriented medicine, too, we are familiar with opposites in the organism that we refer to in terms of veins and arteries, sympathetic and parasympathetic nerves, warmth and cold, or in anthroposophic medicine, the polarities of nerves and blood, synthesis and breakdown, or inflammation and sclerosis.

Now how should we regard the effect of acupuncture — which, like homeopathy, works with the finest energies, but as an external application solely via the system of skin, nerves and senses?

We know from modern research that nerves and blood vessels come closer to the surface at acupuncture points, and that during puncture more hormones such as endomorphine and cortisone are released into the blood. It is also known in the West that in certain regions of the skin the inner organs are reflected in the so-called reflex or Head's zones. Puncturing causes a small shock that releases constricted energies at the relevant place, thus enabling them to flow again. Whoever has experienced whole-body acupuncture or even only ear acupuncture will easily understand this.

Here I'd like to mention an interesting study on three groups of patients suffering from pain, contrasting classic Chinese acupuncture with orthodox medicine's standard pain-alleviating measures such as analgesics, physiotherapy and massage.[77] The procedure was tested on a large number of patients with pain caused by chronic back and knee arthritis. The orthodox therapies performed nearly 50 per cent worse than Chinese treatment using acupuncture. The amazing thing about these results was that in the third group of patients an 'apparent acupuncture' procedure was carried out — that is, instead of puncturing at classic meridian points, needles were inserted at 'slightly wrong' places. Yet these results proved much better than orthodox medical measures. Naturally this gave both sides much to think about, throwing up questions about suggestibility, scientific rigour and verifiability.

The results in the case of the 'slightly misplaced' puncturing could be connected with the fact that schools of acupuncture in China, Korea and Japan, cite acupuncture points that are not the same as each other. People have been trying for a long time to reach some kind of firm agreement about the point sites between the various schools. The success obtained by puncturing 'near' but not on the classical Chinese point might be through touching on a Japanese or Korean point.

Even though many riddles remain to be answered, we can't get away with claiming the outcome was 'only' due to a placebo effect. It is obvious that every prick in the skin releases analgesic messenger substances. We have no doubt all experienced how pinching ourselves in a different place from the pain can momentarily alleviate the latter.

In the same way that allopathic medicine has its failures, and much that is empirically used still remains to be scientifically proven, we should grant this also to Chinese medicine and homeopathy, as important complementary and experiential sciences.

Why not, therefore, draw insights and inspiration from Chinese medicine, even if we don't use every aspect of it?

I myself have profited by it a great deal. In patients suffering from skin complaints — localized inflammation, tumours or eczema — I have looked where the respective meridian was running to see whether it could indicate which underlying organ was sick. I remember a patient, a lady, who kept on getting oozing boils on the temples, and gained no relief from the tried and tested homeopathic remedy for this condition. In Chinese medicine the gall bladder meridian runs along the temples. This raised for me the question of certain gall bladder symptoms that I had overlooked. The proven plant remedy for gall bladder disorders, chelidonium, cured these external symptoms permanently. It remains a question for me, however, whether the unbelievably bitter medicinal herbs and sometimes exotic compounds of Chinese medicine (wormwood is almost sweet and tasty by comparison!) are suitable for everyone in the West.

What is certain is that Chinese medicine is founded on the principle of 'patterns of disharmony', predicated on the idea of a 'weft without a weaver'. The weft or fabric itself is the macro-cosm, the universe that, according to the Chinese view, was not created, but exists through the force of its innermost nature, i.e. through the constant unfolding of the forces of yin and yang. Thus there is also no transcendent 'truth' behind or above things and phenomena, nor a creator or *primum mobile*. The 'truth' can at most be described poetically – but then in a wonderful way! For the Chinese, truth is immanent in everything, and is the perceived process itself.

We see here how very divergent are the Chinese and the anthroposophic, spiritual-scientific view of the cosmos and human beings, and to what extent anthroposophically extended medicine relates the human being, nature and illness to a hidden spirituality that flows through human beings and the world, and can be grasped by means of an enhanced process of cognition.

Before we describe some perspectives in anthroposophically-oriented medicine, I want to illustrate the Chinese view of illness with the example of a stomach ulcer.

In the western medical view, all patients with the same diagnosis are suffering from the same illness – in this case a stomach ulcer. In Chinese medicine, as in homeopathy, the person's whole constitution also plays an important part.

The Chinese doctor establishes for instance the fact that the pains get worse when he touches the patient, but a cold compress alleviates them. The patient appears very robust in constitution, has a reddish complexion and a full, deep voice, and certain aggressive features in his character. He suffers from constipation, the colour of his urine is dark yellow, his tongue has a yellow coating like fat. His pulse is 'full' and tends to be 'tense like wire'. Here the Chinese doctor would diagnose the disharmony pattern 'damp heat, which has affected the spleen'.

But in another patient with the same western diagnosis, the Chinese doctor will ascertain a quite different pattern of disharmony. This patient, let us say, is thin, his complexion more ashen-grey, his cheeks reddish, his palms sweaty, he has a strong thirst, and insomnia with associated nevousness. His tongue is dry and not coated, the pulse 'fine', and a little too fast. The

Chinese doctor will describe this as a 'deficiency of yin that impairs the stomach'. All I want to show here is how essential it is to acquaint oneself with the Chinese way of thinking in order to understand all this and make use of it.

Let us now pass on to a few basic principles of anthroposophically extended medicine as Rudolf Steiner established it in collaboration with physicians.

Anthroposophically extended medicine, which is not in any conflict whatever with mainstream medicine—the doctors that represent it have all undergone modern medical training—is based on Rudolf Steiner's research, as represented in many lectures and writings on epistemology and anthroposophic medical insights into the human being. One of Steiner's primary ideas for understanding the physical, psychological and spiritual organization of a healthy or sick person is the 'threefold' organism, which, by his own testimony, he studied for over 30 years. Some of these aspects have already been presented in the course of this book. The study of threefolding forms the basis for understanding processes in the upper, middle and lower realms or systems of the human being, and their harmony and disharmony, so that if need be these can be rebalanced by suitable measures. If a so-called neurosensory process is too dominant in its catabolic activity—here, too, it is not only a matter of anatomical structures, but of functions—then we see processes of sickness that tend towards sclerotic conditions, the so-called 'cold' illnesses. If on the other hand metabolic activities dominate, with their dissolving tendency, then inflammatory illnesses arise. For this reason some modern pathologists describe 'inflammations' as 'parenteral digestion' or as ' intensified misplaced metabolic activity'. This gives rise to an interesting polarity between inflammation and sclerosis, or tumour formation. Thus raised temperature becomes the 'enemy' of every kind of tendency to hardening or cancer in the organism. Pathogens that occur with inflammation are not true 'pathogens' as such but at the most indicators making a nuisance of themselves in an environment that has already suffered harm of some kind; but they can of course cause secondary damage.

Taking a Chinese view of illness, we might also speak here of a 'pattern of disharmony' in polarities of the 'upper and the lower system' that have fallen out of functional balance, and need to be rebalanced by appropriate measures. In anthroposophic medicine, unlike Chinese medicine, this involves drawing—right into the field of pedagogy—on the third, or rhythmic system. The idea of the threefold organism sheds light on diverse other medical applications, like cupping, crystal therapies, Bach flower remedies, or homeopathic potentization. Thus various external applications—compresses, baths, colour therapy, acupuncture, massages, aromatherapy—speak predominantly to the neurosensory system, intravenous injections to the rhythmic system of the blood, and the medicines taken internally, or movement exercises, to the metabolic limb system.

The different homeopathic potentization stages also become more comprehensible in this light, in that, in the case of acute illnesses, we work on metabolism through lower potencies or mother tinctures, intervene with medium potencies in the rhythmic configuration where life forces and soul forces 'meet', and apply higher or high potencies in the upper, neurosensory organization that must be addressed especially in the case of chronic or psychological illnesses.

Now what are called the four members of our being play a part in this fundamental cosmic pattern from which the individual physical, psychological and spiritual human being is constituted: physical body, life or etheric body, soul or astral body, and the I as the spiritual body, the eternal core of being, also called the entelechy. To understand its nature and mode of action in the natural and the human realm is intrinsic to the lifelong reflections and observations of an anthroposophic physician. Seen purely in physical terms, these four bodily members have an effect on mineral structures (physical body), the realm of life such as regeneration and growth in the glandular and lymphatic system (life body), on conscious and subconscious nerve structures as the material basis of our soul life (astral body), and on the activities of the blood as the bearer of the ego or I. To understand their individual interplay in the functions of the individual organs and in the soul and to see this in connection with natural substances and pharmaceutical processes is one of the main emphases of anthroposophical diagnosis of the 'supersensible sheaths'. Thus the life or etheric body, as a self-enclosed unity, is not only active especially during the night in upbuilding and regenerative processes, but also in memory activity, which is, after all, dependant on the life forces at its disposal, and in which, as in the body, structures are protected from 'decay' or forgetting. Just as a somatic memory is at work in regeneration, so there is likewise a memory capacity in the soul. In death, this life body withdraws, and the body passes over into its chemical elements.

Given, say, a patient who is pale, has weak powers of regeneration, has a weak memory, assimilates food poorly, wakes up tired in the morning, suffers from cold, falls ill or catches an infection easily—someone, therefore, suffering from depleted psychological and organic vitality—the anthroposophic physician will say they have an etheric weakness, which can naturally also originate in the psyche.

From this perspective of our threefold and fourfold nature, there are interesting connections with the kingdoms of nature. We find threefoldness also in the plant kingdom, which is the purest expression of life forces in rhythmic connection with the cosmos. We see it in the functional lawfulness of blossom, leaves and root, which however is the inverse of the human being's threefold organism: blossom corresponds to metabolism, leaves to the rhythmic functions, and the root to the neurosensory system. Homeopathic applications drawn from these three areas are based on this perception: arnica root, for instance, in the case of a head injury, bryony root for congestions, calmus root for rickets, thus formatively; blossoms like camomile, lime, elder, night-blooming cereus (Cactus grandiflorus) have a beneficial effect on the metabolic process, for instance of the heart; and the gigantic leaves of the butterbur (Petasites) support the lungs. Naturally this must not be understood as a fixed schema, since a great many combinations are also possible, and in a certain respect it also depends on the substance in the plants.

In this way a human being is seen as a microcosmic configuration in the macrocosm, an idea familiar to us from Hildegard von Bingen and Paracelsus. Although there are frequently similarities with knowledge from ancient and traditional sources, the anthroposophic path of research must be recognized as innovative in extending thinking into the life of feeling and will.

For instance in his first course for doctors in 1920 Rudolf Steiner begins with polarity in that he shows 'correspondences' between a cough (lungs) and diarrhoea (intestines), one of the correspondences or 'orientations' similar to those we know of from Chinese medicine, though Steiner did not draw it from that source.[78] The heart as the organ of rhythm then lies between them in a mediating capacity.

The anthroposophic doctor starts from the exact phenomenology of corporeality and natural processing, i.e. exact observation, and, through the tangible symptomatology endeavours to grasp the inner workings of the life body, soul body and I. Rudolf Steiner calls this process of knowledge 'intuitive', because it embraces our whole soul and spiritual nature, that is, the realm of feeling and will.[79]

According to Steiner's view, our interior self-perception, our whole inner 'indwelling' in the I, is the 'model' for intuitive knowledge. So pictures of the world, including symptoms of illness, are not the thing in itself but only the outer expression through which something specific is speaking. Only when the essential nature of the things perceived itself lives in the soul of the beholder can one speak of 'intuition', true indwelling or identification with things. Steiner's key statement about this is:

> Perception of one's own 'I' is the model for all intuitive knowledge. To enter into things in this way one has first of all to step out of oneself. One has to become 'selfless', in order to become fully one with the 'self', the 'I', of another entity.[80]

How differently does a doctor connect with a plant that he has been studying for months and that he has learnt thoroughly to love, and that he can rely on—such as the wonderful arnica—

than with a chemical formula of which there is no perception and that means nothing to him!

In case histories that tell us more about Steiner's mode of observation and recommendations, we possess a wealth of suggestions available to interested or practising doctors, so that they can link their own observations and therapeutic approaches to spiritual-scientific knowledge, and deepen their medical insights. For instance, how do diverse substances affect the configuration of the supersensible sheaths, and how can I learn from symptoms how and where the spiritual forces engage? To conclude, let me quote from Steiner's last medical publication with physician Ita Wegman, where they cite a 'characteristic case history' that to me seems to offer useful affirmation of what has been described above.

> A 26-year-old female patient, in an altogether weak and unstable condition. The patient demonstrates that the part of her organism which we refer to in this volume as the astral body, is in a state of excess activity. One sees that her astral body can only be very unsatisfactorily controlled by the ego organization. If the patient decides to do something, the astral body immediately becomes stirred up and agitated. The ego organization tries to prevail but is immediately repulsed. This causes the temperature to rise. The regular activity of digestion is in an eminent sense dependant on a person's normal ego organization. The powerlessness of this ego organization expresses itself in the patient in persistent constipation. This disturbed digestive activity then results in migraine-like conditions and vomiting from which she suffers. During sleep it is apparent that the powerless ego organization causes impaired organic activity in an upward direction from below, with a negative effect on exhalation. The result is an excessive accumulation of carbon dioxide in the organism during sleep, which becomes visible organically in palpitations when waking up, and physically in fear and outcries. Physical examination cannot show anything other than a deficiency of the forces that bring about the proper connection of astral body and physical body. The excessive activity of the astral body means that too little of its forces flow through into the physical and etheric body. The latter will therefore remain frail during their developmental growth. This was

also clear from physical examination of the patient in that she had a delicate and weak body and complained of frequent backache. This comes about because the ego organization has to assert itself most in spinal cord activity. The patient also spoke of having many dreams. That is because the astral body, when it is separated in sleep from the physical and etheric body, unfolds its own activity to excess. So we have to conclude that the ego organization has to be strengthened and the activity of the astral body reduced. In the first case one has to choose a remedy suitable for supporting the ego organization that acts weakly in the digestive tract. Copper is such a remedy. If we apply it in the form of a copper ointment compress placed on the loins the copper will have a strengthening effect on the deficient warmth issuing from the ego organization. We shall notice the effect of this in a diminution of abnormal heart activity and the fading of feelings of fear. The astral body's own excessive activity can be treated by the smallest doses of lead taken internally. Lead contracts the astral body and awakens in it the forces by which it connects more strongly with the physical and the etheric body.[*] The patient visibly improved under this treatment. A certain steadiness and self-confidence replaced her unstable condition. The patient's frame of mind changed from distress to inner satisfaction. Symptoms of constipation and back pain disappeared, as did the migraine-like conditions and headaches. The patient once again became able to work.[81]

---

[*] Lead poisoning consists of too strong a connection of the astral with the etheric and physical body, so that these two bodies become subject to an over-pronounced catabolic process.

Gleaned on the Journey

'A human being's heart signifies heat or the element of fire and is indeed heat, for heat in the whole body has its origin in the heart. The bladder signifies the element of air, and the air also rules within it. The liver signifies the element of water and is also water; for it is from the liver that blood enters the whole body and all our bodily members; the liver is the mother of the blood.

The lungs signify the earth and are also of the same quality.

So the earth, too, would produce no fruit and no metal would grow in it, neither gold nor silver, copper, iron nor stone, if the stars did not act within it. Nor would any grasses grow out of it without the action of the stars. The head signifies the heavens; it has grown upon the body with its veins and channels of force, and all our strength comes from our head and brain into the body, into the vessels that are the fount of flesh.

You see, a human being has within him bile, which is poison, and he cannot live without it, for the bile makes the sidereal spirits mobile, joyful, triumphant, or full of laughter, for it is a source of joys. But where it inflames itself in one element, it spoils the whole human being for anger in the sidereal spirits comes from the bile.

That is: if the bile rises up and flows to the heart, it kindles the element of fire, and the fire kindles the sidereal spirits that reign in the blood of the veins in the element of water; then the whole body trembles in anger and the poison of the bile. Joy has the same source and is of the same substance as anger. Thus when the bile is kindled in a loving or sweet quality, in that which is pleasant for a person, then the whole body trembles with joy, which sometimes affects the sidereal spirits also, if the bile rises up too much and is kindled in the sweet quality.'[82]

'Just as the doctors who take ill people under their wings to look after them have to know about what a human being, life and health are, and in which way a balance, a harmony of the parts is to be maintained or an imbalance is able to ruin or destroy them, so each one who has a good knowledge of these elements can better supply a remedy than someone who hasn't. [...] The same is also necessary for a misconstrued cathedral, in so far as a physician-architect must understand what a building is, and out of which principles a correctly built building is constructed and how many parts it can be divided into and what the causes are that hold the building together and make it lasting, and what the nature of gravity is and what the potency of strength, and in what way it can be put together and combined and what effects these can show together. Whoever has true knowledge of these ideas will present the result to your satisfaction.'[83]

'Every activity has to be practised through movement.
Knowing and willing are two human activities.
Distinguishing, judging and counselling are human actions.
Our body is subject to the heavens and the heavens are subject to the spirit.'[84]

'After I had for months only been considering the spirit, my body, which could no longer stand this situation, reached for the most extreme measures to claim its rights: I became very ill [...] In its way it was not an uninteresting time. It is an odd thing to find oneself less an active person than a stage, the scenario where microbes are waging their battles. And then at times of physical weakness, one experiences psychological shifts, which were not unwelcome to me as a novelty. During the illness traits of my being appeared which were usually hidden; the feminine aspect acquired the upper hand, showing the world in a different, more personal and friendlier light. During such times I am lacking in will, lacking in wishes, and remember my habitual exertions, often so forcefully undertaken, with the gently smiling sympathy with which a woman looks at the incomprehensible ambition of men.

Now I am convalescent and I always enjoy this condition immensely. Otherwise I usually sense my body as something foreign to me, as material given up unrealizably to the spirit, with no inner connection to myself. The spirit is now behaving quite passively, while the regenerating physical forces are ruling all the more industriously; and the consciousness centred in the body has a cheering feeling of lasting productivity.

The happy feeling of a small child is probably like this. A grown-up has states of similar contentment only during bodily weakness, and to a lesser and lesser degree the more he becomes intellectual. The theoretically normal psycho-physical balance, where the centre of consciousness is placed between the physical and the psyche, so that both of them appear real to the same degree, is not a natural condition to anyone like us, and can never be. However different the dimensions are to which the life of the body and the spirit belong—it is *one* energy that is expended in both spheres, and where it is to be adequate to the highest demands in *one* sphere, the other one must correspondingly be neglected [...].

Indeed, it does us good once in a while to exist purely as a body, to do nothing but just to let things happen to us. Such periods also signify a natural reaction to times of increased spiritual exertion [...].

Thus someone who seeks to achieve something here on earth, will never overpower nature—whose normal path, however, never goes straight but in a spiral form. The alternation of various states of consciousness, the rhythmic alternation of interests, is in the same sense necessary and salutary as the alternation between waking and sleeping. I have long become unaccustomed to suffer from periods of depression, or to be appalled at times of stultification: I know that temporary stultification is definitely the precondition for future enlightenment.'[85]

'On 13 December 1795, Wieland received an invitation to visit Goethe. But Wieland, who had a high temperature the next night, had to cancel his visit.

In reply Goethe sent him some medicine consisting of goose liver in jelly.

On December 16 Wieland thanked him as follows:

Dear Brother,

When your tasty medicine arrived yesterday I was just starting to recover after a high temperature, which took a lot out of me on Monday night. In spite of the strong belief I have in you, as you know, I doubted very much whether goose liver and jelly was a very beneficial choice in my circumstances: But my physician (Dr Huschke) who happened to come along, crossed and blessed himself when he heard that you had sent us goose liver to cure me of an indisposition, whatever it was. His opinion was that this either signified an attack on my life (which he did not suspect), an *ignorantia vincibilis* (easily surmountable ignorance) of the first rank [...], or a *tertium quoddam* (a certain third thing) which was beyond his or any professional determination [...].

Besides, the storm of fever had no particular consequences, and all I am waiting for (provided that your medicine keeps that long) is for my stomach to be restored again *in integrum* so that, despite all professional advice, I can eat it and wish you good health as I do, whatever the effect may be on my own.'[86]

'Smoking,' says Goethe, 'makes us stupid; it makes us incapable of thinking and writing poetry. Also it is only for idle people, for people who are bored, who sleep for a third of their lives, spend a third of their lives eating, drinking and doing other necessary and unnecessary things, and then, although they keep on saying that life is short, do not know what they shall do with the last third of life. For such lazy rascals a comforting association with a pipe and the comfortable view of a cloud of smoke that they blow into the air is an amusing entertainment because it helps to pass the hours. Smoking brings with it beer drinking, so that the heated gums are cooled again. Beer thickens the blood, and intoxication is at the same time intensified by the narcotic tobacco smoke. So the nerves are deadened and the blood thickened to the point of faltering. If this were to continue, as it seems likely to do, then after two or three generations we shall see what these beer bellies and puffing louts have made of Germany. It will first become apparent in the stupidity, deformity and impoverishment of our literature, and yet these fellows will greatly admire such wretchedness. And at what cost is this horror! Twenty-five million thalers already go up in tobacco smoke in Germany. The amount could rise to 40, 50, 60 million. And this will feed no hungry person or clothe anyone's nakedness. Just think what could be done with this money! But inherent to smoking is a terrible rudeness too, an anti-social impertinence. Smokers taint the air far and wide and suffocate every honest man who is not able to smoke to defend himself. Who is capable of entering a smoker's room without feeling nauseated? Who can stay there without perishing?'[87]

'The wrappings of the outer form hide the inner organs. We investigate these inner organs in physiology, in biology, according to their form, their structure. There is no other option for us if we find ourselves, to begin with, on the foundation of the natural science customary today. But in reality the lungs, stomach, heart, liver and kidneys, as human organs, are not what they appear to be when we look at them in their enclosed form, in what I would like to call their largely immobile structure, especially immobile to human sensory observation. No, this form is deceptive, for in the living human being these individual organs are in continuous living movement. They are by no means organs formed in an immovable state, but living processes, and we should actually not speak of lungs, heart, kidneys and liver. We should speak of a heart process, a sum of heart processes, a sum of lung processes, a sum of kidney processes; for what is taking place here is a continuous metamorphosis whose self-enclosure is only such that the whole organism can — and indeed for *outer* observation must — be seen as one, whole form. However, to advance from considering form that actually only reveals the outer aspect, to a living process, to what really changes at every moment in those organs, to what actually constitutes the life process of these organs, is impossible with sensory observation. It requires mobile inner vision as this exists in imaginative knowledge [...].

Human beings frequently believe that materialism can be overcome by just leaving the whole world of matter to itself out there in the world, by taking leave of it spiritually, and raising themselves into a spiritually abstract existence in cloud-cuckoo-land, where they indulge in mystic fumblings, regarding the material world as a lower world over which one should elevate oneself [...].

What spiritual science is trying to do is to penetrate everything with the spirit with which it has imbued itself, to tell us what spiritual essence and spiritual reality lives in each human organ: what the nature of the lungs is, the nature of the liver, the heart, the stomach, and so on, when looked at spiritually, how the spirit and the soul penetrate the whole human organism; it seeks to shine the light of spirit right into the smallest cells, so that nothing remains that is not illumined by the light of the spirit. Then we no longer have matter on the one hand and abstract spirit on the other; then there grows together and unites in reality the abstract spirit on the one hand, and abstract matter on the other.'[88]

'Infirmity alone makes us take notice and learn, and enables us to analyse processes which we would otherwise know nothing about. A man who falls straight into bed every night, and ceases to live until the moment when he wakes and rises, will surely never dream of making, not necessarily great discoveries, but even minor observations about sleep. He scarcely knows that he is asleep. A little insomnia is not without its value in making us appreciate sleep, in throwing a ray of light upon that darkness. An unfailing memory is not a very powerful incentive to study the phenomena of memory.'[89]

'Griefs, at the moment when they change into ideas, lose some of their power to injure our heart.'[90]

'If suffering is a bad thing, then which part of your being is it bad for? If it is bad for the body, then let the body complain; and if it is bad for the spirit, then the spirit has the power to withhold the cause of pain and suffering. No pain can penetrate that part of your being that is gifted with reason.'[91]

'Anyone who is afraid of death is afraid either of the cessation of every feeling, or he is afraid of feeling different from the way he feels in earthly life. But if you do not feel anything any more — then there is neither evil nor suffering; but if, rather, your perceptions become different, all this means is that a person's nature is changing, but not that he stops living.'[92]

Notes and references

1. Sandor Marai, *Himmel und Erde. Betrachtungen*, Munich 2001.
2. F. Nietzsche, in *Wie man wird, was man ist. Ermutigung zum kritischen Denken*, ed. by Ursula Michels-Wenz, Frankfurt a/M 1988.
3. Sherwin Nuland, *The Mysteries Within: A Surgeon Explores Myth, Medicine, and the Human Body*, New York, Simon & Schuster, 2000.
4. Sherwin Nuland, op. cit.
5. See Jörg Blech, *Die Krankheitserfinder. Wie wir zu Patienten gemacht werden*, Frankfurt a/M 2003.
6. From Wolfgang G.A. Schmidt, *Der Klassiker des Gelben Kaisers zur inneren Medizin*, Freiburg im Breisgau 1993.
7. Afterword in Guy de Maupassant, *Weitere fünfzig Novellen*. Selected and translated into German by N.O. Scarpi, Zurich 1990.
8. Leonardo da Vinci, *Die Aphorismen, Rätsel und Prophezeiungen*. Selected and translated into German by Marianne Schneider, Munich 2003.
9. Mark Hertsgaard, *Im Schatten des Sternenbanner, Amerika und der Rest der Welt*, Munich 2003.
10. *Die Erschaffung der Menschen*, from *Bulgarische Märchen*, ed. by Elena Ognjanwa, Frankfurt a/M 1992.
11. Leonardo da Vinci, op. cit.
12. Friedrich Nietzsche, from *Wie man wird, was man ist*, op. cit.
13. Leonardo da Vinci, op. cit.
14. Rudolf Steiner, *Mythen und Sagen. Okkulte Zeichen und Symbole*, GA [Collected Works] 101, lecture of 15 September 1907. See *Occult Signs and Symbols*, Anthroposophic Press, 1972.
15. Michael Gershon, *Der kluge Bauch. Die Entdeckung des zweiten Gehirns*, Munich 2001.
16. *Bulgarische Märchen*, op. cit., *Die Erschaffung des Teufels*.
17. E.W. Heine, *Wenn Steine sprechen*, from *Kille Kille. Makabre Geschichten*, Zurich 2002.
18. Michael Frensch, *Unterscheidung der Geister anhand der Apokalypse des Johannes*, Schaffhausen 2004.
19. Friedrich Nietzsche, *Ecce Homo. How One Becomes What One Is*, Penguin Classics 1992.
20. *The New Testament* as translated by Emil Bock, Stuttgart 1983.

21. Rudolf Steiner, *Der Mensch in Zusammenhang mit dem Kosmos 9: Nordische und mitteleuropäische Geistimpulse*, GA 209, lecture of 24 November 1921. See *Self-Consciousness*, Garber Communications, 1986.

22. Thomas Moore, *The Planets Within. The astrological psychology of Marsilio Ficino*, Lindisfarne Press, New York 1990.

23. See Arthur Schult, *Mysterienweisheit im deutschen Volksmärchen*, Bietigheim 1980.

24. Sherwin Nuland, op. cit.

25. Rudolf Steiner, *Geisteswissenschaft und Medizin*, GA 312, lecture of 5 April 1920. (English translation available as *Introducing Anthroposophical Medicine*, Anthroposophic Press 1999.)

26. Ibid.

27. Rüdiger Safranski, *Schiller: oder die Erfindung des deutschen Idealismus*, Frankfurt a/M 2015.

28. Lewis Thomas, *The Lives of a Cell: Notes of a Biology Watcher*, Penguin 1978.

29. Rudolf Steiner, *Heilpädagogischer Kurs*, GA 317, lecture of 25 June 1924. (English translation available as: *Education for Special Needs*, Rudolf Steiner Press 2014.)

30. Ibid.

31. Rudolf Steiner, *Mensch und Welt*, GA 351, lecture of 13 October 1923.

32. William Harvey, *De motu cordis et sanguinis in animalibus*, 1628, quoted in Fritz-Heinz Holler, *Analogien in der Geschichte der Kreislaufslehre. Anthroposophisch-medizinisches Jahrbuch*, vol. 2, Stuttgart 1951.

33. Elisabeth Wellendorf, *Mit dem Herzen eines anderen leben. Die seelischen Folgen der Organtransplantation*, Zurich 1998. See also Paolo Bavastro, 'Herztransplantation — eine kritische Betrachtung', in *Das Herz des Menschen*, ed. by Paolo Bavastro and Hans Christoph Kümmel, Stuttgart 1999.

34. 'Die drei Soldaten und der Doktor', in *Märchen von Krankheit und Heilung*, edited with an Afterword by Stephan Marks, Frankfurt a/M 1996.

35. See Johannes W. Rohen, 'Organe des rhythmischen Systems', in *Morphologie des menschlichen Organismus*, Stuttgart 2002.

36. Leon Manteuffel-Szoege, *Über die Bewegung des Blutes. Hämodynamische Untersuchungen*, Stuttgart 1977.

37. 'Der Richter', in *Chinesische Geister – und Liebesgeschichten*, selected by Martin Buber, Zurich 1948.
38. Thomas Fuchs, *Die Mechanisierung des Herzens. Harvey und Descartes – Der vitale und mechanische Aspekt des Kreislaufs*, Frankfurt a/M 1992.
39. Thomas Fuchs, 'Herzgeschichte', *NZZ Folio*, the *Neue Zürcher Zeitung* magazine.
40. Willibald Gawlik, *Arzneimittelbild und Persönlichkeitsporträt, Konstitutionsmittel in der Homöopathie*, Stuttgart 1990.
41. Wolfgang G.A. Schmidt, *Der Klassiker des Gelben Kaisers zur inneren Medizin*, op. cit.
42. Alla Selawry, *Metall-Funktionstypen in Psychologie und Medizin*, Heidelberg 1985.
43. See also Ernst-Michael Kranich, 'Veränderungen der Persönlichkeit bei Niereninsuffizienz', in *Der innere mensch und sein Leib. Eine Anthropologie*, Stuttgart 2003.
44. In Goethe's letter to Wilhelm von Humboldt.
45. 'Grosse Abhandlung über die Harmonie der Vier Jahreszeiten mit dem menschlichen Geist', in Wolfgang G.A. Schmidt, *Der Klassiker des Gelben Kaisers zur inneren Medizin*, op. cit.
46. Rudolf Steiner, GA 312, op. cit., lecture of 29 March 1920.
47. Heinrich Huebschmann, *Psyche und Tuberkulose. Beiträge aus der allgemeinen Medizin*, a series edited by Professor Viktor von Weizsäcker MD, issue 8, Stuttgart 1958.
48. H.E. Richter, *Die Chance des Gewissens*, Hamburg 1986.
49. Ida Cermak, *Ich klage nicht. Begegnungen mit der Krankheit in Selbstzeugnissen schöpferischer Menschen*, Zurich 1983.
50. George Sand, *Geschichte meines Lebens*, selections from her autobiography with an introduction by Renate Wiggershaus, Frankfurt a/M 1994.
51. Rudolf Steiner, 'Notes to the first medical course' in *Beiträge zur Rudolf Steiner-Gesamtausgabe*, Dornach 1971.
52. Herbalist and priest Johann Künzle, *Chrut und Uchrut*, Minusio 1988.
53. Wolfgang G.A. Schmidt, *Der Klassiker des Gelben Kaisers zur inneren Medizin*, op. cit.
54. Rudolf Steiner, GA 312, op. cit., see note 25.
55. Rudolf Steiner, *Die Erkenntnis des Übersinnlichen in unserer Zeit und deren Bedeutung für das heutige Leben*, GA 350, lecture of 6 June 1923. See *From Mammoths to Mediums*, Rudolf Steiner Press, 2000.

56. Hermann Laubeck, 'Die Neubewertung der Physiologie der Herz—und Blutbewegung', in *Das Herz des Menschen*, ed. by Paolo Bavastro and Hans Christoph Kümmel, Stuttgart 1999.

57. See *Das Herz stärken. Ganzheitliche Selbsthilfe bei Infarkt und Herzschwäche*, ed. by A. Bopp, Th. Breitkreuz, A. Fried, J. Gruber, Munich 2001.

58. See also: Eberhard J. Wormer, *Strophantin*. Rottenburg 2015.

59. See note 46 above.

60. See Olaf Koob, *Die kranke Haut*, Stuttgart 2002.

61. Herman Melville, *The Confidence-Man: His Masquerade*, Oxford World's Classics 2008.

62. See André Dufour, 'Geschichte der Urologie', in *Illustrierte Geschichte der Medizin*, vol. 3, Salzburg 1990.

63. See note 45 above.

64. G. Suchantke, 'Über den Zusammenhang des Seelisch-Geistigen im Menschen mit seiner Leibesnatur' in *Natura — eine Zeitschrift zur Erweiterung der Heilkunst nach geisteswissenschaftlicher Menschenkunde*, 1929/30, reprint, Arlesheim 1981.

65. Rudolf Steiner, *Natur und Mensch in geisteswissenschaftlicher Betrachtung*, GA 352, lecture of 23 February 1924. (English translation available as *From Elephants to Einstein*, Rudolf Steiner Press 1998.)

66. Rudolf Steiner/Ita Wegman, *Grundlegendes für eine Erweiterung der Heilkunst nach geisteswissenschaftlichen Erkenntnissen*, GA 27. (English edition available as *Extending Practical Medicine*, Rudolf Steiner Press 1996.)

67. Sun Ssemiao, Chinese physician in the sixth century AD, in Lin Yutang, *Weisheit des lächelnden Lebens*, Frankfurt a/M and Leipzig 2004.

68. In *Ein Leben für den Geist. Ehrenfried Pfeiffer (1899–1961)*. Edited and introduced by Thomas Meyer, Basel 1999.

69. Heinrich Schipperges, *Am Leitfaden des Leibs. Zur Anthropologik und Therapeutik Friedrich Nietzsches*, Stuttgart 1975.

70. Hans-Ulrich Grimm, *Die Ernährungslüge. Wie uns die Lebensmittelindustrie um den Verstand bringt*, Munich 2003.

71. Rudolf Steiner, *Aus den Inhalten der esoterischen Stunden, Gedächtnisaufzeichnungen von Teilnehmern. Band 1, 1904–1909*, GA 266/1. (English edition available as *From the Contents of the Esoteric Lessons: Sketches from Memory by Participants. Volume 1*, SteinerBooks 2011.)

72. Rudolf Steiner, *Geisteswissenschaftliche Grundlagen zum Gedeihen der Landwirtschaft. Landwirtschaftlicher Kursus*, GA 327. (English edition available as *Agriculture Course*, Rudolf Steiner Press 1974.)
73. Treatise by a royal physician at the Mongolian court, in Lin Yutang, op. cit.
74. Rudolf Steiner, *Die Schöpfung der Welt und des Menschen, Erdenleben und Sternenwirken*, GA 354, lecture of 12 July 1923. (English version available as *From Sunspots to Strawberries*, Rudolf Steiner Press 2002.)
75. In Ruth Jensen, *Das Herz ist ein Affe. Dasein und Wahrheit im Verständnis der Kulturen Chinas und des Abendlands*, Schaffhausen 1997.
76. Ted J. Kaptchuk, *Das grosse Buch der chinesischen Medizin. Die Medizin von Yin und Yang in Theorie und Praxis*, Munich 2001.
77. *Der Spiegel* magazine, 44/2004.
78. Rudolf Steiner, GA 312, op. cit.
79. See also Volker Fintelmann, *Intuitive Medizin. Einführung in eine anthroposophisch ergänzte Medizin*, Stuttgart 1995.
80. Rudolf Steiner, *Die Stufen der höheren Erkenntnis*, GA 12. See *Stages of Higher Knowledge*, Anthroposophic Press, 1981.
81. Rudolf Steiner/Ita Wegman, op. cit.
82. Jakob Böhme, *Im Zeichen der Lilie. Aus den Werken des christlichen Mystikers*, selected by Gerhard Wehr, Munich 1998.
83. Leonardo da Vinci, 'Codex Atlanticus', in Michael White, *Leonardo da Vinci: The First Scientist*, Abacus 2001.
84. Leonardo da Vinci, *Philosophische Tagebücher*, compiled and edited by Giuseppe Zamboni, Hamburg 1958.
85. Count Hermann Keyserling, *Das Reisetagebuch eines Philosophen. Der kürzeste Weg zu sich selbst führt um die Welt herum*, vol. 2, Darmstadt 1921.
86. *Ergetzen ist der Musen erste Pflicht*. Stories and anecdotes about Christoph Martin Wieland, collected and edited by Egon Freitag, Jena 2001.
87. Heinrich Luden, 'Rückblicke in mein Leben (1847)', in *Goethe und die Medizin*, edited by Manfred Wenzel, Frankfurt a/M 1992.
88. Rudolf Steiner, *Anthroposophie, ihre Erkenntniswurzeln und Lebensfrüchte*, GA 78, lecture of 6 September 1921. (English edition available as *The Fruits of Anthroposophy*, Rudolf Steiner Press 1986.)
89. Marcel Proust, from *À la Recherche du Temps Perdu*, in Alain de Botton, *How Proust Can Change Your Life*, Picador 2006.
90. Marcel Proust, op. cit.

91. Marcus Aurelius, *Wie soll man leben? Anton Čechov liest Marc Aurel*, edited, translated from the Russian and with a Foreword by Peter Urban, Zurich 2001.
92. Marcus Aurelius, op. cit.

Further Reading

Anthroposophic Medicine for all the Family, Recognizing and treating the most common disorders, Sergio Maria Francardo, Rudolf Steiner Press, 2017

An Introduction to Anthroposophic Medicine, Extending the Art of Healing, Victor Bott, Rudolf Steiner Press, 2004

Living With Your Body, Health, Illness and Understanding the Human Being, Walther Bühler, Rudolf Steiner Press, 2013

By Rudolf Steiner:

Disease, Karma and Healing, Spiritual-Scientific Enquiries into the Nature of the Human Being, Rudolf Steiner Press, 2013

Extending Practical Medicine, Fundamental Principles Based on the Science of the Spirit (with Dr Ita Wegman), Rudolf Steiner Press, 2000

Good Health, Self-Education and the Secret of Well-being, Rudolf Steiner Press, 2017

The Healing Process, Spirit, Nature and Our Bodies, Rudolf Steiner Press, 2011

Introducing Anthroposophical Medicine, SteinerBooks, 2011

Nutrition, Food, Health and Spiritual Development, Rudolf Steiner Press, 2008

Understanding Healing, Meditative Reflections on Deepening Medicine through Spiritual Science, Rudolf Steiner Press, 2013

A note from the publisher

For more than a quarter of a century, **Temple Lodge Publishing** has made available new thought, ideas and research in the field of spiritual science.

Anthroposophy, as founded by Rudolf Steiner (1861-1925), is commonly known today through its practical applications, principally in education (Steiner-Waldorf schools) and agriculture (biodynamic food and wine). But behind this outer activity stands the core discipline of spiritual science, which continues to be developed and updated. True science can never be static and anthroposophy is living knowledge.

Our list features some of the best contemporary spiritual-scientific work available today, as well as introductory titles. So, visit us online at **www.templelodge.com** and join our emailing list for news on new titles.

If you feel like supporting our work, you can do so by buying our books or making a direct donation (we are a non-profit/ charitable organisation).

office@templelodge.com

TEMPLE LODGE

For the finest books of Science and Spirit